T0176582

Advanced Ceramic Coatings and Materials for Extreme Environments

A Collection of Papers Presented at the 35th International Conference on Advanced Ceramics and Composites January 23–28, 2011 Daytona Beach, Florida

Edited by
Dongming Zhu
Hua-Tay Lin
Yanchun Zhou

Volume Editors
Sujanto Widjaja
Dileep Singh

A John Wiley & Sons, Inc., Publication

Published by John Wiley & Sons, Inc., Hoboken, New Jersey.
Published simultaneously in Canada.

For general information on our other products and services or for technical support, please contact our Customer Care Department within the United States at (800) 762-2974, outside the United States at (317) 572-3993 or fax (317) 572-4002.

Wiley also publishes its books in a variety of electronic formats. Some content that appears in print may not be available in electronic formats. For more information about Wiley products, visit our web site at www.wiley.com.

Library of Congress Cataloging-in-Publication Data is available.

ISBN 978-1-118-05988-3

oBook ISBN: 978-1-118-09523-2
ePDF ISBN: 978-1-118-17258-2

ISSN: 0196-6219

Printed in the United States of America.

10 9 8 7 6 5 4 3 2 1

Contents

**ENVIRONMENTAL BARRIER COATINGS FOR TURBINE
ENGINES AND EXTREME ENVIRONMENTS**

FUNCTIONALLY GRADED COATINGS AND INTERFACES

THERMAL BARRIER COATINGS

**MATERIALS FOR EXTREME ENVIRONMENTS: ULTRA HIGH
TEMPERATURE CERAMICS (UHTCS) AND NANOLAMINATED
TERNARY CARBIDES AND NITRIDES (MAX PHASES)**

Preface

The Symposium on Advanced Ceramic Coatings for Structural, Environmental and Functional Applications and the Materials for Extreme Environments: Ultrahigh Temperature Ceramics and Nanolaminated Ternary Carbides and Nitrides (MAX Phases) Symposium were held at the 35th International Conference on Advanced Ceramics and Composites in Daytona Beach, Florida, January 23–28, 2011. The 16 papers presented in this issue include papers from these two symposia.

We are greatly in debt to the members of the symposium organizing committees, for their assistance in developing and organizing this vibrant and cutting-edge symposium. We also would like to express our sincere thanks to manuscript authors and reviewers, all the symposium participants and session chairs for their contributions to a successful meeting. Finally, we are also grateful to the staff of The American Ceramic Society for their efforts in ensuring an enjoyable conference and the high-quality publication of the proceeding volume.

DONGMING ZHU, *NASA Glenn Research Center, USA*
H. T. LIN, *Oak Ridge National Laboratory, USA*
YANCHUN ZHOU, *Institute of Metal Research, Chinese Academy of Sciences, CHINA*

Introduction

This CESP issue represents papers that were submitted and approved for the proceedings of the 35th International Conference on Advanced Ceramics and Composites (ICACC), held January 23-28, 2011 in Daytona Beach, Florida. ICACC is the most prominent international meeting in the area of advanced structural, functional, and nanoscopic ceramics, composites, and other emerging ceramic materials and technologies. This prestigious conference has been organized by The American Ceramic Society's (ACerS) Engineering Ceramics Division (ECD) since 1977.

The conference was organized into the following symposia and focused sessions:

Symposium 1	Mechanical Behavior and Performance of Ceramics and Composites
Symposium 2	Advanced Ceramic Coatings for Structural, Environmental, and Functional Applications
Symposium 3	8th International Symposium on Solid Oxide Fuel Cells (SOFC): Materials, Science, and Technology
Symposium 4	Armor Ceramics
Symposium 5	Next Generation Bioceramics
Symposium 6	International Symposium on Ceramics for Electric Energy Generation, Storage, and Distribution
Symposium 7	5th International Symposium on Nanostructured Materials and Nanocomposites: Development and Applications
Symposium 8	5th International Symposium on Advanced Processing & Manufacturing Technologies (APMT) for Structural & Multifunctional Materials and Systems
Symposium 9	Porous Ceramics: Novel Developments and Applications

Symposium 10	Thermal Management Materials and Technologies
Symposium 11	Advanced Sensor Technology, Developments and Applications
Symposium 12	Materials for Extreme Environments: Ultrahigh Temperature Ceramics (UHTCs) and Nanolaminated Ternary Carbides and Nitrides (MAX Phases)
Symposium 13	Advanced Ceramics and Composites for Nuclear and Fusion Applications
Symposium 14	Advanced Materials and Technologies for Rechargeable Batteries
Focused Session 1	Geopolymers and other Inorganic Polymers
Focused Session 2	Computational Design, Modeling, Simulation and Characterization of Ceramics and Composites
Special Session	Pacific Rim Engineering Ceramics Summit

The conference proceedings are published into 9 issues of the 2011 Ceramic Engineering & Science Proceedings (CESP); Volume 32, Issues 2-10, 2011 as outlined below:

- Mechanical Properties and Performance of Engineering Ceramics and Composites VI, CESP Volume 32, Issue 2 (includes papers from Symposium 1)
- Advanced Ceramic Coatings and Materials for Extreme Environments, Volume 32, Issue 3 (includes papers from Symposia 2 and 12)
- Advances in Solid Oxide Fuel Cells VI, CESP Volume 32, Issue 4 (includes papers from Symposium 3)
- Advances in Ceramic Armor VII, CESP Volume 32, Issue 5 (includes papers from Symposium 4)
- Advances in Bioceramics and Porous Ceramics IV, CESP Volume 32, Issue 6 (includes papers from Symposia 5 and 9)
- Nanostructured Materials and Nanotechnology V, CESP Volume 32, Issue 7 (includes papers from Symposium 7)
- Advanced Processing and Manufacturing Technologies for Structural and Multifunctional Materials V, CESP Volume 32, Issue 8 (includes papers from Symposium 8)
- Ceramic Materials for Energy Applications, CESP Volume 32, Issue 9 (includes papers from Symposia 6, 13, and 14)
- Developments in Strategic Materials and Computational Design II, CESP Volume 32, Issue 10 (includes papers from Symposium 10 and 11 and from Focused Sessions 1, and 2)

The organization of the Daytona Beach meeting and the publication of these proceedings were possible thanks to the professional staff of ACerS and the tireless dedication of many ECD members. We would especially like to express our sincere thanks to the symposia organizers, session chairs, presenters and conference attendees, for their efforts and enthusiastic participation in the vibrant and cutting-edge conference.

ACerS and the ECD invite you to attend the 36th International Conference on Advanced Ceramics and Composites (http://www.ceramics.org/daytona2012) January 22-27, 2012 in Daytona Beach, Florida.

SUJANTO WIDJAJA AND DILEEP SINGH
Volume Editors
June 2011

Advanced Coating Characterization Methods and Non-Destructive Evaluation

MONITORING DELAMINATION OF THERMAL BARRIER COATINGS DURING INTERRUPTED HIGH-HEAT-FLUX LASER TESTING USING UPCONVERSION LUMINESCENCE IMAGING

Jeffrey I. Eldridge and Dongming Zhu
NASA Glenn Research Center
Cleveland, OH 44135

Douglas E. Wolfe
Applied Research Laboratory
The Pennsylvania State University
University Park, PA 16802

ABSTRACT

Upconversion luminescence imaging of thermal barrier coatings (TBCs) has been shown to successfully monitor TBC delamination progression during interrupted furnace cycling. However, furnace cycling does not adequately model engine conditions where TBC-coated components are subjected to significant heat fluxes that produce through-thickness temperature gradients that may alter both the rate and path of delamination progression. Therefore, new measurements are presented based on luminescence imaging of TBC-coated specimens subjected to interrupted high-heat-flux laser cycling exposures that much better simulate the thermal gradients present in engine conditions. The TBCs tested were deposited by electron-beam physical vapor deposition (EB-PVD) and were composed of 7wt% yttria-stabilized zirconia (7YSZ) with an integrated delamination sensing layer composed of 7YSZ co-doped with erbium and ytterbium (7YSZ:Er,Yb). The high-heat-flux exposures that produce the desired through-thickness thermal gradients were performed using a high power CO_2 laser operating at a wavelength of 10.6 microns. Upconversion luminescence images revealed the debond progression produced by the cyclic high-heat-flux exposures and these results were compared to that observed for furnace cycling.

INTRODUCTION

The demonstration of nondestructive diagnostic tools for monitoring delamination progression for TBC-coated specimens due to thermal cycling has been mostly limited to exposure to furnace environments, where thermal cycling alternates between hot and cold isothermal conditions. However, furnace cycling does not adequately simulate turbine engine conditions, where TBC-coated components are subjected to significant heat fluxes that are not present in a furnace environment. These heat fluxes produce through-thickness temperature gradients that may alter both the rate and pathway of TBC delamination progression. In contrast to the non-evolving temperature profiles associated with furnace cycling, temperature gradients through TBCs subjected to high heat fluxes will also change over the course of cycling due to TBC sintering and the generation of delamination cracks, both of which will affect heat transport through the TBC. Therefore, diagnostics developed to predict remaining TBC life in engine environments must be based on testing of TBCs subjected to engine-like heat flux conditions.

Luminescence imaging of TBCs incorporating an integrated delamination sensing layer has been shown to provide exceptionally high contrast for monitoring TBC delamination.[1,2] The

3

objective of this paper is to extend luminescence-based delamination monitoring to TBCs subjected to high heat flux. To meet this objective, upconversion luminescence imaging has been applied to monitor TBC delamination progression in TBC-coated specimens subjected to high heat fluxes produced by a high-power CO_2 laser. The delamination progression revealed by upconversion luminescence imaging for two different heat fluxes is compared with the delamination progression produced by furnace cycling.

EXPERIMENTAL PROCEDURES
Upconversion Luminescence Imaging
        Upconversion luminescence refers to the special case of luminescence where the emission wavelength, λ_{em}, is shorter (higher energy) than the excitation wavelength, λ_{ex}. Upconversion luminescence is attractive for delamination monitoring because of the absence of background emission at wavelengths shorter than the excitation wavelength, resulting in superior contrast. Fig. 1 shows the concept and the coating design behind monitoring TBC delamination by upconversion luminescence imaging. The 7YSZ TBC incorporates a thin base layer that is co-doped with erbium and ytterbium (YSZ:Er,Yb) below a thicker undoped YSZ layer. Erbium was selected as a dopant specifically because a two-photon excitation (λ_{ex} = 980 nm) from the $^4I_{15/2}$ ground state of Er^{3+} to the $^4F_{7/2}$ excited state can produce upconversion luminescence emission at λ_{em} = 562 nm, corresponding to the relaxation from the $^4S_{3/2}$ excited state back to the $^4I_{15/2}$ ground state of Er^{3+}.[4] The Yb^{3+} co-dopant is added because Yb^{3+} is a better absorber of the 980 nm excitation (via excitation from the Yb^{3+} $^2F_{7/2}$ ground state to the $^2F_{5/2}$ excited state) than Er^{3+} and can then transfer the excitation energy to Er^{3+} to produce luminescence. The delamination contrast that is desired is achieved because when delamination cracks form near the bottom of the TBC, these cracks introduce interfaces that are highly reflective for both the excitation and emission wavelengths so that significantly higher luminescence intensity is observed from regions containing delamination cracks.[1,2]

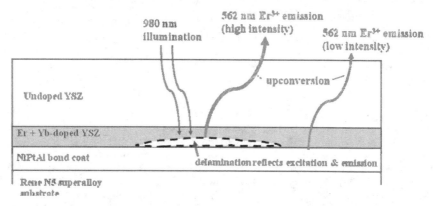

Figure 1. Concept for monitoring TBC delamination progression by upconversion luminescence.

The excitation source for the upconversion luminescence was a 980 nm laser diode that illuminated the specimen after the laser beam traveled through beam expanding optics. The upconversions luminescence images were collected by a cooled CCD camera with a bandpass filter centered at 562 nm (FWHM = 40 nm) and an image acquisition time of 3.25 sec. A background subtraction was performed for each upconversion luminescence image followed by normalization to an unconversion luminescence image of an uncycled control specimen.

Specimens

TBCs were deposited by multiple ingot electron-beam physical vapor deposition (EB-PVD) at Penn State and consisted of an initial 11 μm Er+Yb co-doped YSZ layer (cation dopant mole concentrations of 1% Er and 3% Yb) followed by a 135 μm thick undoped 7YSZ overlayer which was deposited with no disruption of the columnar growth between layers in a manner previously described.[1-3] The TBCs were deposited onto NiPtAl bond-coated (Chromalloy) nickel-based superalloy Rene N5 substrates (25.4 mm diameter, 3.18 mm thick).

High-Heat-Flux Laser Testing

TBC-coated specimens were subjected to heat fluxes using the NASA GRC high-heat-flux laser facility.[5] In short, a combination of high power CO_2 laser (wavelength = 10.6 μm) heating of the surface of the TBC combined with forced air backside cooling produced the desired heat flux. The heat flux through the specimen was determined by subtracting heat flux losses by reflection and radiation from the heat flux delivered by the laser. TBC surface temperatures were determined by an 8 μm pyrometer and the substrate backside temperatures by a two-color pyrometer. As previously shown,[5] the apparent thermal conductivity of the TBC, k_{TBC}, could be determined from the heat flux, top and backside temperatures, and known substrate thermal conductivity. For the laser cyclic testing, one cycle consisted of 60 min with the laser on followed by 3 min with the laser off.

Two heat flux conditions were examined. A lower heat flux of 95 W/cm^2 produced nominal surface and TBC/bond coat interface temperatures of 1290°C and 1140°C, respectively, while a higher heat flux of 125 W/cm^2 produced nominal surface and interface temperatures of 1345°C and 1175°C, respectively. As shown in Fig. 2 for the higher heat flux test, actual temperatures fluctuated somewhat from these nominal temperatures, and there was a drift upward in the TBC surface temperature, especially toward the end of cyclic life. Upconversion luminescence images were collected during interruptions in laser cycling (typically after 20 cycle intervals). Results from the laser testing were compared with conventional furnace cycling in a tube furnace where each cycle consisted of 45 min at 1163°C followed by 15 min cooling (to ~120°C) and upconversion luminescence images were collected at interruptions in furnace cycling (also typically after 20 cycle intervals). All laser and furnace cycle testing was performed on specimens with TBCs deposited in the same EB-PVD run. One specimen was cycled to failure at each of the heat flux conditions and two specimens were cycled to failure by furnace cycling.

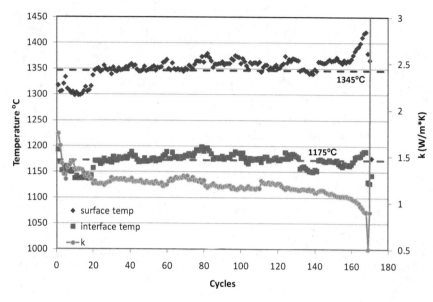

Figure 2. TBC surface (red diamond), interface (blue square) temperatures and apparent k_{TBC} (green circles) as a function of laser cycles for the 125 W/cm^2 heat flux testing.

RESULTS

A selected subset of the upconversion luminescence images collected during furnace cycling and for interrupted laser cycle testing at 95 and 125 W/cm^2 heat flux conditions in Figs. 3, 4, and 5, respectively.

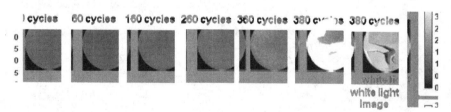

Figure 3. Upconversion luminescence images collected during interrupted furnace cycling to 1163°C along with white light image of final specimen failure at 380 furnace cycles. Normalized intensity scale indicates ratio of intensity to that of an uncycled control specimen.

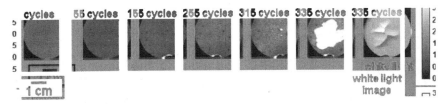

Figure 4. Upconversion luminescence images collected during interrupted laser cycling with heat flux of 95 W/cm^2 along with white light image of final specimen failure at 335 cycles. Normalized intensity scale indicates ratio of intensity to that of an uncycled control specimen.

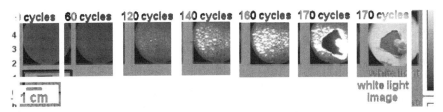

Figure 5. Upconversion luminescence images collected during interrupted laser cycling with heat flux of 125 W/cm^2 along with white light image of final specimen failure at 170 cycles. Normalized intensity scale indicates ratio of intensity to that of an uncycled control specimen.

In all cases, an initial reduction in luminescence intensity is observed followed by a gradual, but consistent increase in luminescence intensity that is spotty in nature, although the increase in luminescence is less spotty for the furnace cycling test. Cycling was continued until macroscopic failure was observed (see white light images in Figs. 3, 4, and 5), consisting of TBC buckling after 380 furnace cycles, TBC buckling for the 95 W/cm^2 heat flux after 335 laser cycles, and TBC spallation for the 125 W/cm^2 heat flux after 170 laser cycles. Note that the bright spots in the luminescence images, corresponding to local TBC/substrate separations, cover a much higher fraction of the image area prior to macroscopic failure for the higher 125W/cm^2 heat flux.

While the lower heat flux (95 W/cm^2) cycling and the furnace cycling showed no evidence of delamination progression to the naked eye until final TBC macroscopic failure, this was not the case for the higher heat flux (125 W/cm^2) cycling where undulations of the TBC surface were evident under glancing illumination well before final macroscopic TBC failure (Fig. 6). It is clear that the raised areas of the TBC surface correspond to the bright areas in the upconversion luminescence image, indicating that the raised areas correspond to regions where there is a local TBC/substrate separation. This surface texturing has been observed previously,[6,7] and has been attributed to bond coat rumpling beneath the TBC.

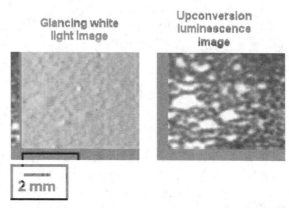

Figure 6. Higher magnification images of region of TBC-coated specimens after 155 laser cycles with heat flux of 125 W/cm². (Left) White light image under glancing illumination. (Right) Upconversion luminescence image of same area.

The changes in upconversion luminescence intensity can be compared more quantitatively by plotting the normalized luminescence intensity averages over the entire area of each specimen image as a function of cycles as shown in Fig. 7 for the 95 and 125 W/cm² heat flux laser cycling tests as well as the two 1163°C furnace cycling tests. All the plots show several features in common: (1) an initial decrease in luminescence intensity until minimum is reach (marked by triangles in Fig. 7) followed by (2) a gradual increase in intensity associated with debond progression and (3) a substantial inflection upward (marked by circles in Fig. 7) associated with macroscopic TBC failure. In addition, both furnace cycling tests and the 95 W/cm² heat flux laser cycling test exhibited the onset of macroscopic TBC failure (marked by circles in Fig. 7) at about the same normalized luminescence intensity. Significant differences included the 95 W/cm² laser cycling test exhibiting an initial incubation period where significant luminescence increase was not observed until beyond 100 cycles, followed by an accelerating increase in luminescence intensity. The behavior for the 125 W/cm² laser cycling test exhibited pronounced differences in comparison to the lower heat flux laser cycling or furnace cycling: not only was the rate of luminescence increase faster (not unexpected in view of the higher interface temperatures), but the average normalized luminescence increased to much higher intensity before the onset of macroscopic TBC failure.

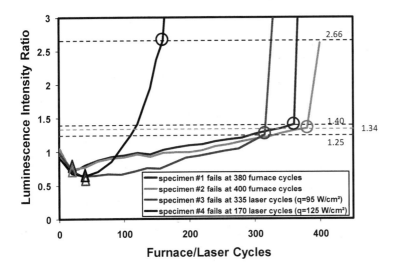

Figure 7. Normalized upconversion luminescence intensity averages over entire specimen image as a function of cycles. Comparison in behavior for furnace cycling (at 1163°C), and two conditions of high-heat-flux laser cycling (q = 95 and 125 W/cm²). Circles mark first observation of macroscopic TBC failure. Triangles mark minimum luminescence.

DISCUSSION

Upconversion luminescence imaging was used to compare TBC delamination progression due to thermal cycling under two laser-induced heat flux conditions with delamination progression produced by furnace cycling. In all cases, an initial reduction in upconversion luminescence intensity was observed (Fig. 7) that is most likely associated with growth of the thermally grown oxide (TGO) beneath the TBC resulting in a less reflective substrate. The bright spots in the luminescence images (Figs. 3-6) are associated with local delamination cracks that act as highly reflective interfaces for both the $\lambda_{ex} = 980$ nm excitation and $\lambda_{ex} = 562$ nm upconversion emission. The growth in number, size, and intensity of these bright spots is proposed to correspond to the generation, lateral growth, and increasing crack opening, respectively, of delamination cracks. Preliminary SEM inspection of previously furnace cycled TBC-coated specimens has verified the presence of isolated local delaminations. SEM investigation of the cross-sections of the laser and furnace cycled TBC-coated specimens tested for this paper have not yet been performed but are planned.

While the general trend of debond progression by generation of delamination cracks that grow in number and extent and coalesce until macroscopic TBC failure is observed in all cases (Figs. 3-6), significant differences were also observed. Some of the observed differences may be attributed to differences in the interface temperatures during cycling, which were 1140°C and 1175°C for the 95 W/cm² and 125 W/cm² heat flux laser cycling, respectively, and an intermediate 1163°C for the furnace cycling tests. Some differences could also be due to the shorter hot cycle duration (45 min vs. 60 min) for the furnace cycle tests, and the faster heating and cooling associated with the laser cycling tests. To determine the effects of heat flux on delamination progression, furnace cycling should be performed at temperatures that match the interface temperatures during the heat flux laser tests (furnace cycling at 1140°C and 1175°C) and with the same hot time duration (60 min), and this is planned for future work. However, the present results still suggest effects of heat flux on delamination progression as described below.

While the onset of macroscopic failure intensity occurred at about the same normalized luminescence intensity for the lower 95 W/cm^2 heat flux laser cycling and the furnace cycling tests (Fig. 7), the 95 W/cm^2 laser cycling exhibits an incubation period until about 100 laser cycles before the increases in luminescence intensity associated with delamination progression are observed. This incubation (or latent) period is followed by an accelerating increase in luminescence intensity in comparison to the furnace cycling tests. Although only one laser cycling test was performed at this heat flux, this behavior was very out-of-character in comparison to the furnace cycling tests, where this behavior was not observed not only from the 2 furnace cycling tests from the same TBC deposition batch (Fig. 7), but also never observed in furnace cycling under the same conditions for 16 other specimens from other TBC deposition batches. These results suggest that the 95 W/cm^2 heat flux does not accelerate delamination initiation. However, it is proposed that once delamination cracks are established, the heat flux significantly accelerates delamination progression. It is important, therefore, to realize that the equivalent interface damage accumulation appears to occur much later in the laser cycling than in the furnace cycling (damage accumulation proceeds more uniformly during furnace cycling, but is weighted more heavily towards the end of life during laser cycling). Therefore, a prediction of TBC remaining life based on interface damage accumulation during furnace cycling would grossly overestimate remaining life when heat flux effects are present.

The higher 125 W/cm^2 laser cycling test does not exhibit the same prolonged incubation period as the lower 95 W/cm^2 test; however, onset of delamination progression (luminescence intensity increase marked by black triangle in Fig. 7) is still delayed in comparison with the furnace cycling tests despite the higher interface temperature attained in the 125 W/cm^2 laser cycling (1175° C vs. 1163°C). This comparison suggests that the high heat flux may actually suppress delamination initiation. However, following delamination initiation, the rate of luminescence increase (delamination damage accumulation) is greatly accelerated for the higher heat flux laser cycling test. Perhaps the most striking difference observed for the higher heat flux laser cycling test is that the delamination damage associated with bright luminescence (Figs. 5-7) accumulates to a much greater extent before macroscopic TBC failure occurs, suggesting that under the conditions of higher heat flux and higher interface and surface temperatures, the TBC can accommodate a much higher degree of damage before producing macroscopic TBC buckling or spallation. This very different behavior may be related to a different mechanism driving delamination progression. In particular, TBC surface undulations (Fig. 6) associated with bond coat rumpling beneath the TBC were only observed for the higher heat flux condition. The luminescence images clearly indicate that the raised portions of the TBC surface (corresponding to peaks in the surface of the rumpled bond coat below) are associated with local areas of TBC/substrate separation; therefore, delamination progression in the higher heat flux testing appears to be driven by bond coat rumpling. The much greater role of bond coat rumpling in the delamination progression in the higher heat flux test may be partly due to the higher interface temperature that will promote bond coat rumpling because of higher rates of diffusion in the bond coat, but may also be due to the heat flux contribution which results in higher TBC temperatures above the interface that will reduce the stiffness of the TBC (along with higher TBC creep rates) and therefore reduce the constraint to bond coat rumpling that the TBC provides.[6,7] Because bond coat rumpling allows a local lengthening of the TGO, rumpling will therefore significantly reduce strain energy near the interface so that the stress intensity at a crack tip along the rumpled bond coat surface will be significantly less than in the absence of bond coat rumpling. This rumpling-induced reduction of crack tip stress intensity will make local TBC buckling more stable against further growth and therefore explain the accommodation of greater delamination damage accumulation prior to macroscopic failure when bond coat rumpling occurs.

TBC delamination produced by thermal cycling has been shown to be driven by TGO growth, mismatch strain energy produced during cooling cycles,[8,9] along with effects of bond coat rumpling[6,7] and effects of TBC sintering.[10] These delamination driving forces can all be affected by heat flux effects as described below.

While TGO growth should only be dependent on interface temperature irrespective of heat flux, delamination cracks can lead to increased temperature at delamination crack tips. For a sufficient crack opening, a delamination crack in the presence of a heat flux will experience a temperature difference, ΔT_{crack}, across the crack, and therefore the temperature above the crack will be greater than at regions away from the crack. Therefore, the temperature at the crack tip will be intermediate between the temperature just above the crack and the temperature at a lateral distance away from the crack, the crack tip temperature will be higher than at distances away from the crack and therefore produce faster TGO growth at the crack tip.

In addition to stresses produced by TGO growth, Evans and Hutchinson[9] have shown that the steady state linear elastic energy release rate for delamination at the TBC/substrate interface during cooling, including release of forces and moments induced by heat fluxes, is

$$G = \frac{d(1-v_{TBC}^2)}{6E_{TBC}}(3\overline{\sigma}^2 + \sigma_0^2) \tag{1}$$

where d is the TBC thickness, and E_{TBC} and v_{TBC} are the elastic modulus and the Poisson's ratio of the TBC, respectively, and σ_0 and $\overline{\sigma}$ are given by

$$\sigma_0 = \frac{E_{TBC}\alpha_{TBC}\Delta T_{surf/int}}{2(1-v_{TBC})} \tag{2}$$

$$\overline{\sigma} = \frac{E_{TBC}}{1-v_{TBC}}\left[\frac{\Delta T_{surf/int}}{2} - \Delta\alpha\Delta T_{int}\right] \tag{3}$$

where ΔT_{int} is the temperature drop of the interface from its peak temperature, $\Delta T_{surf/int}$ is the temperature drop of the surface minus the temperature drop of the interface, α_{TBC} is the coefficient of thermal expansion of the TBC and $\Delta\alpha$ is the coefficient of thermal expansion of the substrate minus that of the TBC. Essentially, $\overline{\sigma}$ is the average stress in the TBC and σ_0 is the difference in stress between the TBC surface and interface upon cooling. Note that σ_0 is purely a heat flux effect and is zero in the absence of a heat flux (when $\Delta T_{surf/int} = 0$). On the other hand, the heat flux contribution in equation 3 (first term) acts in opposition to the TBC/substrate mismatch strain and therefore will reduce $\overline{\sigma}$. A priori, it is not known whether the introduction of a heat flux will increase G (equation 1), since that will depend on whether the increase in σ_0 more than offsets the increase in $\overline{\sigma}$. Because the observed effect of a heat flux (Fig. 7) seems to be to modestly delay debond initiation compared to furnace cycling, the results suggest that the heat flux reduces G at the initial stages of crack propagation.

High heat fluxes can also produce sintering at the higher temperatures near the TBC surface and therefore increase k_{TBC} and produce a thermal conductivity gradient across the TBC thickness with highest thermal conductivity at the surface.[10] Note that the *decrease* in apparent k_{TBC} observed in Fig. 2 is due to the effect of delamination cracks producing a barrier to heat transport. The resulting thermal conductivity gradient will produce an associated change in the thermal profile across the TBC. In addition, sintering will result in an increased effective modulus of the TBC, thereby increasing both $\bar{\sigma}$ and σ_0 and therefore G. The effect of the dimensional sintering-induced shrinkage is less unequivocal since this shrinkage will increase σ_0 but decrease $\bar{\sigma}$.

Finally, the effect of heat flux will also be to reduce the effective TBC stiffness at high temperature due to the higher TBC temperatures above the interface (resulting in lower modulus and higher creep rates). This reduced stiffness will lessen the degree to which the TBC constrains bond coat rumpling, therefore promoting bond coat rumpling induced delamination. Bond coat rumpling will accelerate delamination crack accumulation, but will also accommodate a greater damage accumulation before triggering macroscopic TBC failure because of the strain relaxation that accompanies rumpling.

SUMMARY

Upconversion luminescence imaging successfully monitored TBC delamination progression for both interrupted furnace cycling and high-heat-flux laser cycling tests. It appears that the effect of heat flux is to first delay initial TBC delamination progression due to slightly reduced thermal expansion mismatch stresses, but to then greatly accelerate delamination progression after delamination progression is established, presumably due to an evolution of the mechanical properties of the TBC along with associated changes in the thermal profile across its thickness. Heat fluxes can also promote bond coat rumpling by reducing the TBC stiffness constraint against rumpling, thereby changing the character of the delamination progression. It is proposed that rumpling-induced local delamination produces more stress relief than local delamination produced without bond coat rumpling so that rumpling-induced local delaminations are more resistant to crack growth. Finally, life prediction based on observed damage evolution occurring during furnace cycling will grossly overestimate TBC remaining life under high heat flux conditions.

ACKNOWLEDGMENTS

Funding by the NASA Fundamental Aeronautics Program Subsonic Fixed Wing Program is gratefully acknowledged.

REFERENCES

[1]J.I. Eldridge, T.J. Bencic, C.M. Spuckler, J. Singh, and D.E. Wolfe, "Delamination-Indicating Thermal Barrier Coatings Using YSZ:Eu Sublayers," *J. Am.. Ceram. Soc.*, **89**[10], 3246-3251 (2006).

[2]J.I. Eldridge and T.J. Bencic, "Monitoring Delamination of Plasma-Sprayed Thermal Barrier Coatings by Reflectance-Enhanced Luminescence," *Surf. Coat. Technol.*, **201**[7], 3926-3930 (2006).

[3]J.I. Eldridge, J. Singh, and D.E. Wolfe, "Erosion-Indicating Thermal Barrier Coatings Using Luminescent Sublayers, *J. Amer. Ceram. Soc.*, **89**[10], 3252-3254 (2006).

[4] H. Berthou and C.K. Jörgensen, "Optical-Fiber Temperature Sensor Based on Upconversion-Excited Fluorescence," *Optics Lett.*, **15**[19], 1100-1102 (1990).

[5]D. Zhu, R.A. Miller, B.A. Nagaraj, and R.W. Bruce, "Thermal Conductivity of EB-PVD Thermal Barrier Coatings Evaluated by a Steady-State Laser Heat Flux Technique," *Surf. Coat. Technol.*, **138**[1], 1-8 (2001).

[6]V.K. Tolpygo and D.R. Clarke, "Morphological Evolution of Thermal Barrier Coatings Induced by Cyclic Oxidation," *Surf. Coat. Technol.*, **163-164**, 81-86 (2003).

[7]V.K. Tolpygo, D.R. Clarke, and K.S. Murphy, "Evolution of Interface Degradation During Cyclic Oxidation of EB-PVD Thermal Barrier Coatings and Correlation with TGO Luminescence," *Surf. Coat. Technol.*, **188-189**, 62-70 (2004).

[8]J.W. Hutchinson and A.G. Evans, "On the Delamination of Thermal Barrier Coatings in a Thermal Gradient," *Surf. Coat. Technol.*, **149**, 179-184 (2002).

[9]A.G. Evans and J.W. Hutchinson, "The Mechanics of Coating Delamination in Thermal Gradients," *Surf. Coat. Technol.*, **201**, 7905-7916 (2007).

[10]Y. Tan, J.P. Longtin, S. Sampath, and D. Zhu, "Temperature-Gradient Effects in Thermal Barrier Coatings: An Investigation Through Modeling, High Heat Flux Test, and Embedded Sensor," *J. Am. Ceram. Soc.*, **93**[10], 3418-3426 (2010).

THERMAL IMAGING MEASUREMENT ACCURACY FOR THERMAL PROPERTIES OF THERMAL BARRIER COATINGS

J. G. Sun
Argonne National Laboratory
Argonne, IL 60439

Thermal barrier coatings (TBCs) are being extensively used for improving the performance and extending the life of combustor and gas turbine components. TBC thermal properties, thermal conductivity and heat capacity (the product of density and specific heat), are important parameters in those applications. These TBC properties are usually measured by destructive methods, involving separating the ceramic coating layer from the substrate and performing density, specific heat, and thermal diffusivity measurements. Nondestructive evaluation (NDE) methods, on the other hand, allow for direct TBC property measurement on natural TBC samples so they can be used for inspecting the quality of as-processed components as well as monitoring TBC degradation during service. For this purpose, a multilayer thermal-modeling NDE method has been developed which analyzes data obtained from one-sided pulsed thermal imaging to determine thermal conductivity and heat capacity distributions over the entire surface of a TBC specimen. The measurement accuracy can be affected by many factors from experimental and sample condition variations. These factors are discussed and evaluated in this study based on analytical and numerical simulation results for thermal imaging conditions.

INTRODUCTION

Thermal barrier coatings (TBCs) have been extensively used on hot gas-path components in gas turbines. In this application, a thermally insulating ceramic topcoat (the TBC) is bonded to a thin oxidation-resistant metal coating (the bond coat) on a metal substrate. TBC coated components can therefore be operated at higher temperatures, with improved performance and extended lifetime [1,2]. Because TBCs play critical role in protecting the substrate components, their failure (spallation) may lead to unplanned outage or safety threatening conditions. Therefore, it is necessary to determine the initial condition of as-processed TBCs as well as to monitor the condition change during service to assure their quality and reliability. Because the primary function of a TBC is for thermal insulation, the most important TBC parameters are thermal properties, particularly the thermal conductivity. Experimental methods to accurately measure TBC thermal conductivity are still being actively pursued today.

TBC conductivity can be measured by several methods. The most reliable and commonly used method is laser flash method on stand-alone TBC coated specimens [3,4], which is a special case of the two-sided thermal-imaging method. This method however is a destructive method, requiring the TBC coat to be separated from the substrate for the measurement. Alternatively, laser flash test may also be conducted on specially-prepared TBC specimens, including the top coat, bond coat, and substrate. TBC conductivity is then determined based on multilayer analytical solutions or correlations for the TBC material system [5,6].

Laser flash methods for either single- or multiple-layer materials require two-sided access of the specimens, so cannot be used to analyze TBCs coated on real components with variable substrate thickness. One-sided thermal imaging method, on the other hand, may nondestructively measure TBC thermal properties on real TBC-coated components so may be used for inspection and monitoring of TBC conditions during service. One such method is the multilayer thermal modeling method developed at Argonne National Laboratory [7]. This method may determine two TBC thermal properties, typically the thermal conductivity and heat capacity when TBC thickness is known. In contrast, two-sided laser flash methods can usually determine only one TBC thermal property, the thermal diffusivity when TBC thickness is known. Therefore, one-sided method has advantages over two-sided methods in terms of the capability to measure more material properties and on nondestructive evaluation (NDE) of real structural components.

The measurement accuracy for both two-sided and one-sided methods is dependent on many factors related to the experimental system, test procedure, and test-sample conditions. Because the fundamental physical process of the test is the same and the measurement is performed using similar or same instrumentations (the only difference is on which side the data are acquired), the same set of factors affects both one- and two-sided methods, although their significance varies. In this investigation, these factors are identified and discussed first for both methods. Theoretical analyses are then performed to determine the measurement accuracy for one-sided multilayer modeling method. Data generated from analytical solutions and numerical simulations are used in the analyses.

TWO- AND ONE-SIDED PULSED THERMAL IMAGING

Pulsed thermal imaging is based on monitoring the temperature change on a specimen surface after it is applied with a pulsed thermal energy that is gradually transferred inside the specimen. The premise is that the heat transfer from the surface (or surface temperature/time response) is affected by internal material structures and properties. Most thermal imaging system uses a photographic flash lamp (or a laser) to apply pulsed heating on sample surface and an infrared camera to measure surface temperature variation during the transient heat-transfer period. Depending on the arrangement of the flash lamp and the infrared camera, a thermal imaging system is two-sided if the heat source and infrared camera are placed on the opposite sides of the test specimen, or one-sided if both heat source and infrared camera are placed at the same side (see Figs. 1a and 1b). A laser-flash system is a special case of the two-sided thermal imaging, in which a pulsed laser is used as the heat source and a single infrared detector (or a temperature sensing probe) is used to measure the average temperature of the unheated back surface. By analyzing the measured surface temperature/time response, one or more material property of the test specimen can be determined. The characteristics of the acquired data and the analysis methods for two- and one-sided setup are completely different. Typical data and factors affecting data accuracy and characteristics are examined below.

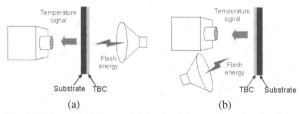

(a)   (b)

Fig. 1. Schematics of (a) two- and (b) one-sided pulsed thermal imaging for 2-layer material.

FACTORS AFFECTING THERMAL IMAGING TESTS

For both one- and two-sided thermal imaging, the fundamental physics for the heat transfer process is the same. A thermal imaging method analyzes the measured temperature transient data on heated or unheated surface to predict material thermal properties, assuming the measured data are accurate. Most thermal imaging methods are based on solutions derived from one-dimensional heat transfer process. These solutions are usually derived based on an ideal condition, i.e., the flash heat is deposited on sample surface (no depth penetration) instantaneously (zero flash duration) and there is no heat loss from all sample surfaces. Solutions for some non-ideal conditions may also be derived and used for data analysis.

The accuracy of the predicted material properties is affected by many factors deviating from the ideal condition. These factors may affect the accuracy of the measured temperature or change the characteristics of the surface temperature transient data (curve) from the ideal ones used by the analysis method. The temperature measurement accuracy is an issue for thermal imaging because an infrared camera only measures the thermal radiation from a surface, not the surface temperature. If the surface is considered a blackbody, i.e., with an emissivity of one, and the entire spectrum of thermal radiation is captured, then radiation intensity is proportional to the fourth power of the surface temperature. However, because an infrared camera is normally sensitive only to a small range of the spectrum, e.g., the mid-infrared range of 3-5 μm, an accurate correlation between the radiation intensity and temperature needs to be determined experimentally. When the surface emissivity is not one, measured radiation intensity is a combination of surface emission and reflection from surrounding background. This further complicates the correlation and its accuracy. Therefore, it is desirable that the sample surface has a higher emissivity so accurate surface temperature can be measured. For TBCs, a carbon-based black paint is usually applied on TBC surface to improve its surface emissivity.

Factors affecting data characteristics may come from experimental system or sample condition. A major factor from the experimental system is the flash duration of the flash lamps, which is typically in the range of 1-5ms [8,9]. Flash duration is a concern for fast thermal transient processes associated with thin and/or high thermal conductivity materials. Factors related to sample conditions include the material translucency (optical transmission), surface roughness, and black surface paint. A translucent material allows the applied flash heat to penetrate inside the material to cause volume heating, which alters the initial condition of the heat transfer process. The surface roughness causes 3D heat conduction near the surface and allows the infrared camera to view temperature at different depths "inside" the material. The black paint, with sufficient thickness and property difference from the TBC material, functions as another layer on the TBC surface. In the following, the effect of these factors for two-sided (laser flash) and one-sided (multilayer modeling) analysis methods are discussed.

DISCUSSION OF FACTORS AFFECTING TWO-SIDED METHODS

From two-sided (laser flash) measurement under ideal conditions, the normalized back-surface temperature transient is illustrated in Fig. 2 as the "ideal curve". The characteristics of the temperature curve include a short time period of no temperature change immediately after the flash, followed by a rapid temperature increase, and eventually approaching a constant. A commonly used method for data analysis is to determine the half-rise time, $t_{1/2}$, which is related only to the sample thickness and thermal diffusivity. Once $t_{1/2}$ is determined from an experimental curve and sample thickness is known, the thermal diffusivity of the material is determined (predicted). However, when the effects of various factors are not negligible, $t_{1/2}$ corrections or other analysis methods are required to predict the thermal diffusivity.

In two-sided thermal imaging or laser flash measurement, the maximum temperature rise on the back surface is usually small, e.g., <3°C. Under such condition, an infrared sensor can be easily calibrated to obtain accurate temperature data. The surface emissivity is also not a problem. This is because that under small temperature changes the sensor reading is linearly related to the temperature and the surface emissivity. Other factors, however, may affect the temperature-time curve. The flash duration, when comparable to the $t_{1/2}$ value, may change the experimental curve drastically so an incorrect $t_{1/2}$ value is derived (see Fig. 2). This problem can be avoided by using a fast flash source (with short duration) and requiring the test sample to have sufficient thickness, or it can be accounted for by theoretical corrections [8]. The translucency of the test material (e.g., TBCs) can be addressed by painting the surface black. This strategy works well for low temperature tests, but is not effective at high temperatures when the radiation heat-transfer component is comparable to the conduction component through the sample thickness. The material translucency will then cause an initial temperature jump, as illustrated in Fig. 2. This problem is best solved by using theoretical models and fitting the model solution with experimental data [4]. Another problem sometimes occurs only in laser flash tests is the heat loss effect, because test samples used in laser flash tests are usually small and mounted by contact fixtures. Heat loss effect has been studied extensively and can be easily corrected [4]. In general, the prediction accuracy of laser flash methods is considered to be within 5% [4].

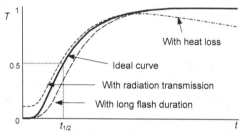

Fig. 2. Typical temperature variations at back surface from two-sided laser flash tests.

ANALYSIS OF MEASUREMENT ACCURACY FOR ONE-SIDED METHOD

In one-sided thermal imaging measurement, the front-surface temperature of a sample is measured. The characteristics of the temperature-time curve for different materials are revealed more clearly when the temperature slope curve is plotted. A TBC material is usually considered a two-layer material in thermal imaging test, because the bond coat is thin and has similar property as the substrate so is considered as part of the substrate. Figure 3 shows a typical temperature slope curve in the log-log scale, identified as "ideal curve", for an opaque two-layer TBC material. Note that the temperature slope, $d(\ln T)/d(\ln t)$, is a nondimensional parameter so its value is unique to a particular sample. The temperature slope is a constant at -0.5 in early times when the flash heat absorbed on TBC surface is propagating only within the top coating layer. When heat approaches the interface, the magnitude of the surface temperature slope increases because the substrate has higher conductivity (so faster heat transfer). The temperature slope eventually approaches to zero when the temperature of the entire sample becomes equalized. The dominant characteristic of the curve is the main peak as indicated in Fig. 3, which has a characteristic position (in time) and magnitude. The multilayer modeling method [7,10] is based on a theoretical solution of the two-layer material and its fitting with the measured curve to determine two parameters, thermal conductivity and heat capacity of the top TBC coating layer (when its thickness is known).

In one-sided experiments, the accuracy of the surface temperature measurement can be difficult to maintain because there is a large temperature variation on the front surface during the experiment. The surface temperature could reach up to 100°C during the thermal flash period, and then decays quickly to <1°C within the majority of the transient measurement period. This requires the infrared camera to have a large dynamic range and an accurate correlation between the measured radiation intensity and temperature over the entire temperature range, so not only the temperature but also the temperature slope is determined accurately. The surface emissivity will also play a significant role to the correlation. In the multilayer modeling method, a dynamic calibration procedure was developed to determine the correct surface temperature and an infrared filter was used to completely eliminate the flash-lamp infrared radiation [11].

Figure 3 illustrates the effects of several factors on the surface temperature slope curve. The material translucency causes optical transmission that may reduce the magnitude of the temperature slope over the entire transient period. This effect is not studied here because it can be easily addressed by applying a black carbon paint on the TBC surface. The flash duration, carbon layer, and surface roughness may modify the temperature slope curve in the early time period, as illustrated in Fig. 3. These three factors, and their effects to measurement accuracy, are analyzed in this investigation. To facilitate the analyses, postulated material properties a two-layer TBC system are assumed and listed in Table 1, where L is layer thickness, k is thermal conductivity, c is heat capacity (where is density and c is specific heat).

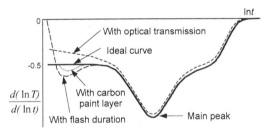

Fig. 3. Typical temperature-slope curves for two-layer TBCs from one-sided thermal imaging.

Table 1. Postulated material properties for TBC material systems used in this study.

Materials	L (mm)	k (W/m-K)	c (J/m³-K)
TBC	0.2	2	$3.016*10^6$
Substrate	1.0	8	$3.237*10^6$
Carbon coat		0.5	$2*10^6$

Flash Duration Effect

Figure 4a shows the flash duration effect on the temperature slope curve. This effect is more sever when flash duration is large, affecting both the position and the magnitude of the main peak. If the flash duration for a data set is unknown and a theoretical model with an arbitrary value, say 1.8ms, is used to analyze (fit) the data, it is obvious that the predicted results are not accurate if the entire curve is used. However, if only part of the curve around the main peak is used in the analysis, e.g., within the range of 10ms to 1s, the results can be improved. This time range for data fitting is used in all analyses presented below. A partial curve fitting is therefore used in this and the following analyses. Figure 4b shows the predicted TBC conductivity and heat capacity when a fixed value of flash duration of 1.8ms is used to analyze the data obtained with different flash durations. The maximum prediction error for conductivity is <1.5% and for heat capacity is <4%. This accuracy is considered good.

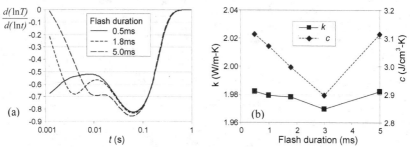

Fig. 4. Flash duration effect on (a) temperature slope data and (b) predicted TBC properties.

Carbon Layer Effect

The carbon paint layer applied on TBC surface for thermal imaging test may have different thermal properties and thicknesses. Figure 5a shows the effect of carbon layers of various thicknesses to the temperature slope curve, based on theoretically generated data with the properties listed in Table 1. Because the thickness of the carbon layer on real samples is usually unknown, it is normally not accounted for in the data analysis. Figure 5b shows the predicted TBC conductivity and heat capacity for TBCs with carbon layers up to 10μm thick (5% of the TBC thickness). The maximum error for conductivity is <4% and for heat capacity is <1% from predictions assuming no carbon layer on surface. This accuracy is also considered good.

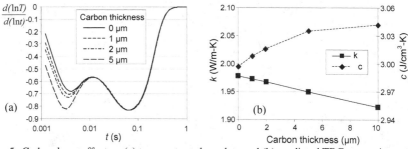

Fig. 5. Carbon layer effect on (a) temperature slope data and (b) predicted TBC properties.

Surface Roughness Effect

The surface roughness effect is analyzed based on 2D numerical simulation data for TBC samples with various surface roughness levels. The simulations were performed using the COMMIX computer code [12], which is a three-dimensional, transient, finite-difference-based code developed and validated for computational heat transfer and fluid flow. The cross-sectional geometry of the TBC sample is illustrated in Fig. 6. Because the roughness is assumed to be periodic and simplified as steps, only the shaded area is used in the simulation. Figure 7a shows the surface temperature slope curves at the peak and valley positions on the TBC surface with a roughness of 10μm and a periodic length of 160μm (flash duration is 0). The surface roughness effect appears mostly in the early times. Although the predicted TBC property values at these positions can be quite different, the average predicted values are very close to the real material values, as shown in Fig. 7b for TBCs with roughness of 10 and 20μm and various periodic lengths. The maximum prediction error for conductivity and for heat capacity is typically <1%. This accuracy is very good.

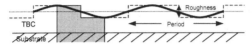

Fig. 6. Cross-sectional geometry of a TBC material with surface roughness.

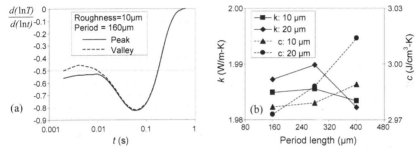

Fig. 7. Surface roughness effect on (a) temperature slope data and (b) predicted TBC properties.

CONCLUSION

Various factors affecting the measurement accuracy of TBC thermal properties from two-sided (laser flash) and one-sided thermal imaging methods were discussed and investigated. These factors either affect the accuracy of the measured surface temperature or change the characteristics of the surface temperature transient curve. For two-sided laser flash methods, most of the factors are either insignificant or can be modeled by well established models. For one-sided method, the multilayer modeling method, those factors may play significant roles on measurement accuracy. In this investigation, three factors relevant to TBC property measurement were studied: the flash duration, carbon-layer thickness, and surface roughness. Analytical solutions and numerical simulations were used to generate the surface temperature data, covering the typical ranges of these factors in TBC measurements. It was identified that these factors all affect the temperature data in the early time period, say <10ms. By proper selection of the data segment used in the analysis by the multilayer modeling method, without modeling the effect of these factors, the error of predicted TBC thermal properties in all cases is <4%. Although further improvement in measurement accuracy may be achieve, these results demonstrate that the multilayer-modeling method is robust and accurate for thermal property measurement as well as NDE characterization for TBCs.

ACKNOWLEDGMENT

This work was sponsored by the U.S. Department of Energy, Office of Fossil Energy, Advanced Research and Technology Development/Materials Program, and by the Heavy Vehicle Propulsion Materials Program, DOE Office of FreedomCAR and Vehicle Technology Program, under contract DE-AC05-00OR22725 with UT-Battelle, LLC.

REFERENCES

1. US National Research Council, National Materials Advisory Board, "Coatings for High Temperature Structural Materials," National Academy Press, Washington, DC, 1996.
2. A. Feuerstein and A. Bolcavage, "Thermal Conductivity of Plasma and EBPVD Thermal Barrier Coatings," Proc. 3rd Int. Surface Engineering Conf., pp. 291-298, 2004.
3. H. Wang and R.B. Dinwiddie, "Reliability of laser flash thermal diffusivity measurements of the thermal barrier coatings," J. Thermal Spray Techno., Vol. 9, pp. 210-214, 2000.
4. E1461-07, "Standard Test Method for Thermal Diffusivity by the Flash Method," ASTM, 2007.
5. B.K. Jang, M. Yoshiya, N. Yamaguchi, and H. Matsubara, "Evaluation of Thermal Conductivity of Zirconia Coating Layers Deposited by EB-PVD," J. Mater. Sci., Vol. 39, pp. 1823-1825, 2004.
6. J.G. Sun, "Thermal Conductivity Measurement for Thermal Barrier Coatings Based on One- and Two-Sided Thermal Imaging Methods," in Review of Quantitative Nondestructive Evaluation, eds. D.O. Thompson and D.E. Chimenti, Vol. 29, pp. 458-469, 2009.
7. J.G. Sun, "Thermal Imaging Characterization of Thermal Barrier Coatings," in Ceramic Eng. Sci. Proc., eds. J. Salem and D. Zhu, Vol.28, no. 3, pp. 53-60, 2007.
8. Sun, J. G., and Erdman, S., "Effect of Finite Flash Duration on Thermal Diffusivity Imaging of High-Diffusivity or Thin Materials," in Review of Quantitative Nondestructive Evaluation, Vol. 23, ed. D.O. Thompson and D.E. Chimenti, pp. 482-497, 2003.
9. J. G. Sun and J. Benz, "Flash Duration Effect in One-Sided Thermal Imaging," in Review of Progress in Quantitative Nondestructive Evaluation, eds. D.O. Thompson and D.E. Chimenti, Vol. 24, pp. 650-654, 2004.
10. J.G. Sun, "Thermal Property Measurement for Thermal Barrier Coatings by Thermal Imaging Method," in Ceramic Eng. Sci. Proc., eds. S. Mathur and T. Ohji, Vol. 31, no. 3, pp. 87-94, 2010.
11. J.G. Sun, "Optical Filter for Flash Lamps in Pulsed Thermal Imaging," U.S. Patent No. 7,538,938 issued May 26, 2009.
12. H. M. Domanus, Y. S. Cha, T. H. Chien, R. C. Schmitt, and W. T. Sha, COMMIX-1C: A Three-Dimensional Transient Single-Phase Computer Program for Thermal-Hydraulic Analysis of Single-Component and Multi-Component Engineering Systems, NUREG/CR-5649, ANL-90/33, Vol. 1, 1990.

Advanced Coating Processing and Nanostructured Coating Systems

HIGH VELOCITY SUSPENSION FLAME SPRAYED (HVSFS) HYDROXYAPATITE COATINGS FOR BIOMEDICAL APPLICATIONS

N. Stiegler, R. Gadow, A. Killinger
Institute for Manufacturing Technologies of Ceramic Components and Composites (IMTCCC)
Universität Stuttgart
Stuttgart, Baden-Württemberg, Germany

ABSTRACT

Thermal spraying of suspensions containing particles of submicron or nano size offers new possibilities in functional coating development and creates new application fields. Spraying nano Hydroxyapatite (HAp) suspensions by means of hypersonic flame spraying (HVSFS), results in coatings with a refined microstructure and a layer thickness typically ranging from 20 - 50 μm can be achieved. HVSFS is a novel thermal spray process developed at IMTCCC, for direct processing of submicron and nanosized particles dispersed in a liquid feedstock.

Thermally sprayed HAp coatings are widely used for various biomedical applications due to the fact that HAp is a bioactive, osteoconductive material capable of forming a direct and firm biological fixation with surrounding bone tissue.

Bioceramic coatings based on nanoscale HAp suspension were thermally sprayed on Ti plates by high-velocity suspension flame spraying (HVSFS) and compared to Atmospheric Plasma Sprayed (APS) as well as High Velocity Oxy Fuel Sprayed (HVOF) ones. Different combustion chamber designs were developed to optimize the HVSFS coating process.

The deposited coatings were mechanically characterized including surface roughness, micro hardness and coating porosity. Phase content and crystallinity of the coatings were evaluated using X-ray diffraction (XRD). The bond strength of the layer composites were analyzed by the pull-off method and compared for different spraying conditions. The coating composite specimen and initial feedstock were further analyzed by Scanning Electron Microscope (SEM) and rheological analysis.

INTRODUCTION

Thermally-sprayed hydroxyapatite (HAp) is widely employed to enhance the osseointegration of medical implants and prostheses in a number of applications, including endoprosthesis such as hip, knee, glenohumeral (shoulder), ginglymus (e.g. elbow) joint or dental implants. Remarkable increases in the number of hip and knee replacement surgeries are predicted. Primary total hip arthroplasty is estimated to grow by 174 percent, from 208,600 in 2005 to 572,100 by 2030, while primary total knee arthroplasty is projected to grow from 450,400 to 3.48 million procedures during the same period (more than 673 percent growth), just in the US [1]. The thermally-sprayed coatings generally consist of hydroxyapatite $[Ca_5(PO_4)_3OH]$, which is chemically similar to the mineral component of human bones and hard tissues [2]. It is indeed able to support bone in-growth and osseointegration when used in orthopedic, dental, and maxillofacial applications [3, 4]. Moreover, HAp coatings have the capacity to shorten the healing process of metal based implants. The molar ratio of the hydroxyapatite in human bone is not stoichiometric (lower than 1.67) and additionally contains sodium, magnesium, carbonate, fluorine and chlorine ions, which are substitutes for Ca^{2+} and PO_4^{3-} in the crystal lattice.

Atmospheric plasma-(APS) and high velocity oxygen-fuel-(HVOF) spraying are well known deposition techniques for bioactive hydroxyapatite coatings [5-11]. The above mentioned processes are limited to conventional dry spray powder with a particle size greater than 5 μm, because fluidization of the powder becomes more challenging with decreasing particle size. Dry feeding of micro-sized or nano-sized particles is only realizable by agglomerated HAp feedstock particles (grain size > 5μm) as performed in [12, 19]. Manufacturing thermal spray coatings from suspension feedstock, instead of conventional dry powder, enables direct delivery of very fine powder particles (micron-sized or nano-

25

sized) into the plasma or gas jet [13-17, 19]. Compared to standard spray powder processing, direct processing of fine particles dispersed in liquid solvent always results in coatings with smaller lamella size [18, 19]. One of the most promising suspension spray technique is the high-velocity suspension flame spraying (HVSFS) process developed at the Institute for Manufacturing Technologies of Ceramic Components and Composites (Universität Stuttgart), proven to run as a stable process [20].

The present research therefore aims to ascertain how different coating processes (APS, HVOF, HVSFS) and process parameters like gas flow, spray distance, air-fuel ratio and electric arc current affect the coating properties (e.g. coating hardness, surface roughness, bond strength) and deposition efficiency.

EXPERIMENTAL

Powder Feedstock and Suspension Characterization
Two different HAp suspensions, supplied by Cer.i.Col Centro Ricerche Colorobbia - Italy, were employed in the HVSFS experiments. Nano-sized HAp particles were dispersed either in water or diethylene glycol. The dry HAp spray powder for the HVOF and APS coating experiments was supplied by CERAM GmbH Ingenieurkeramik - Germany.

The rheological behavior of the suspensions was characterized by a shear rheometer (UDS 200, Anton Paar Germany GmbH) in rotational cylindrical mode. The measured parameters were viscosity η, shear stress τ, and shear rate dv/dy. The measurements were performed at a constant temperature of 20 °C and at different shear rates between 0.1 s^{-1} and 500 s^{-1}.

The grain size distributions of the two suspensions and of the dry powder were studied by laser diffraction particle size analysis and dynamic light scattering analysis. To detect existing agglomeration, the grain size distribution was measured without and with ultrasonic treatment of the supplied suspension and dry powder material. The grain size distribution without ultrasonic treatment was done to detect the size of the spontaneously formed agglomerates representative of the material employed during the coating process. The measurement with ultrasonic treatment was performed in order to study the strength of the agglomerates. Samples of each raw material were investigated by scanning electronic microscopy (SEM), sputter-coated with gold and examined in a LEO VP 438 from Leo Elektronenmikroskopie GmbH, Oberkochen, Germany.

Deposition of the Coatings
The different spray torches were operated on six axis robot systems using a simple meander kinematics with 2 mm offset. The substrates were planar 50×50×3 mm Ti (grade 2) plates, mounted vertically on a sample holder, degreased with acetone and mildly grit blasted using white corundum particles (FEPA 120; Ø = 100 - 125 μm) and 0.5 MPa compressed air, leading to a typical roughness of Ra = 4.5 - 5 μm. The sample was cooled during the deposition by two air nozzles mounted on the torch. The samples were weighed before and after spraying, to determine the coating weight and calculate the deposition efficiency of different processes and parameter sets. The surface temperature during coating process was monitored using a pyrometer type Keller Pz10 AF1.

As described in [21], the feeding system of the HVSFS process equipment draws the suspension out of a reservoir and delivers it continuously under a controlled flow rate axial into the combustion chamber of the HVSFS torch (modified HVOF torch, model: TopGunG, GTV GmbH, Luckenbach, Germany). To prevent sedimentation of the suspension in the reservoir, it was mechanically stirred during the entire coating process. To feed the suspension into the combustion chamber a conical shaped injection nozzle was used. The torch was also equipped with a specially designed combustion chamber. For the HVOF coating experiments a TopGun-G torch from GTV GmbH, Luckenbach, Germany was used. The torch was equipped with a cylindrical combustion chamber. For the APS coating experiments an F6 torch from GTV GmbH, Luckenbach, Germany was used.

Characterization of Deposited Coatings

In order to determine the coating thickness low magnification micrographs were employed. The as-deposited surface roughness was measured by a stylus profilometer (Perthometer PGK, Mahr, Göttingen, Germany: average of four measurements). The porosity of the coatings was measured by digital image analysis on polished cross-sections (Software AQUINTO). SEM micrographs were taken from surface and cross-section of each coating. X-ray diffraction patterns were acquired on the as-deposited coatings and for the pure hydroxyapatite powder for reference. The hardness and elastic modulus were determined according to DIN EN ISO 14577 on the polished cross-sections by microhardness measurement using a Fischerscope H100C equipped with a Vickers indenter. The given values are the average of 10 indentations.

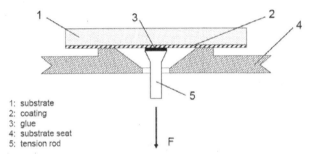

1: substrate
2: coating
3: glue
4: substrate seat
5: tension rod

Figure 1. Schematic illustration of the bond strength measuring method (pull-off).

The bond strength was measured according to ISO 4624 with a miniaturized measuring device on a universal testing machine Zwick Z100 (Fig 1). HTK ULTRA BOND® was used as adhesive. Precipitation heat treatment of the glue took place at 190 °C for 35 minutes. The test speed was set to 0.5 mm/min.

RESULTS AND DISCUSSION

Suspension and Powder Characterization

The suspension properties are a significant aspect for the HVSFS coating process. After axial injection of the liquid feedstock into the combustion chamber of the HVSFS torch, indeed, the suspension jet gets atomized into small droplets by an injection nozzle [22-25]. The liquid phase then evaporates rapidly from the droplets, releasing individual particles or (more frequently) particle agglomerates. Their size strongly influences their thermal history: small particles or small agglomerates rapidly heat up, but they also rapidly cool down resulting in a fairly different behavior concerning the splat formation on the surface [25]. The highest surface temperature during the coating process was measured for the HVSFS-technique between 364 °C and 464 °C. For the APS-technique low surface temperatures between 77 °C and 118 °C were observed, dependent on the process parameters. The surface temperature during HVOF spraying was between 255 °C and 295°C. All of these phenomena are strongly influenced by rheological behavior of the suspension and by the particle agglomeration behavior.

The diethylene glycol suspension shows an almost Newtonian behavior, whereas the water-based one is shear-thinning (Fig. 2): the viscosity decreases with increasing shear rate. Although pure water is a Newtonian fluid, its rheological behavior changes remarkably with addition of solid particles.

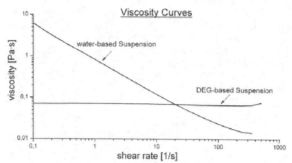

Figure 2. Viscosity curves of the diethylene glycol- and water-based hydroxyapatite suspensions.

Another important point is the stability of the suspension. Indeed, the used suspensions did not show any sedimentation, even after weeks of observation.
The grain size distribution of the two different suspensions for HVSFS and of the dry powder for HVOF and APS are summarized in Fig. 3. The grain size distributive curve for the DEG-based suspension exhibits one peak. This peak is sharp and clearly recognizable, whereas they tend to broaden and merge in the water-based one. The measured grain size distributions of the dry powder used for HVOF and APS coating experiments exhibits one peak, and there is almost no agglomeration of the dry powder particles.

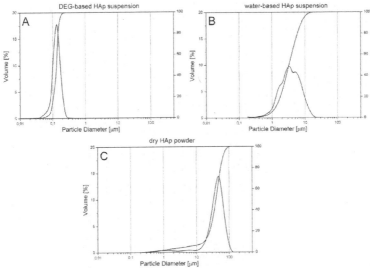

Figure 3. Grain size distributions measured on the diethylene glycol based (A), water-based (B) suspension and dry powder (C).

From SEM micrographs of the two different suspensions it can be stated, that dispersed particles of the two suspensions consist of agglomerated nano-sized (< 50 nm) primary particles. Comparing these very fine nanoparticles to the size distributions provided in Fig. 3 clearly indicates that the hydroxyapatite nanoparticles seem to be strongly cohesive and are mostly agglomerated in both suspensions. The d50 values for the agglomerated particles in the DEG-based and water-based suspension are respectively 135 nm and 3.29 µm.

The dry HAp powder material for the HVOF and APS experiments consists of micrometer-sized particles with d50 = 40 µm. Analysis of the dry powder by SEM images shows a partly blocky, partly splintered grain shape of the crushed material. The particles of the crushed material have a uniformly distributed porosity originating from the sintering process employed to produce the raw material. The raw material consists of small crystallites (with a diameter < 1 µm) formed during the sintering process. After sintering the raw material block is crushed and sieved to the favored grain size. The grains are porous and built up from the above mentioned crystallites.

Structural and Microstructural Properties of the Deposited Coatings

Comparing the different coatings remarkable differences regarding coating thickness and deposition efficiency exist. The HVSFS process has the lowest deposition efficiency in the range of 10 % to 25 %, for the APS process it is in the range of 50 % to 60 % and for the HVOF process the deposition efficiency strongly depends on the process parameters and varies between 40 % and 80 %.

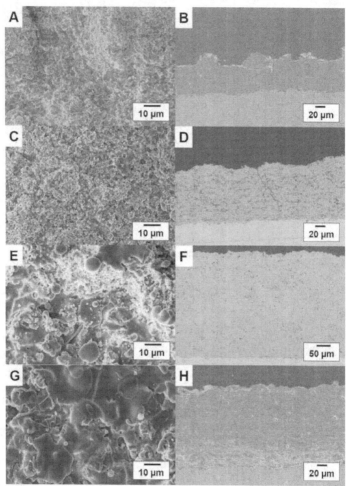

Figure 4. SEM micrographs: (A) Surface of a DEG based HVSFS-HAp coating. (B) Cross-sectional view of a DEG based HVSFS-HAp coating. (C) Surface of a water-based HVSFS-HAp coating. (D) Cross-sectional view of a water-based HVSFS-HAp coating. (E) Surface of a HVOF-HAp coating. (F) Cross-sectional view of a HVOF-HAp coating. (G) Surface of an APS-HAp coating. (H) Cross-sectional view of an APS-HAp coating.

Fig. 4 summarizes the different microstructure and surface topography by means of SEM images. Regarding the HVSFS process the use of diethylene glycol (Fig. 4A, B) or water (Fig. 4C, D) as solvent for the suspension affects the coating properties more than the selection of deposition parameters; indeed, although the latter can have some effect on deposition efficiency and porosity, the two kinds of suspensions produce remarkably different overall microstructural features. The coatings produced by the water-based suspension have a layered structure, i.e. the layers deposited during each of the ten torch cycles performed during the deposition of this coating are clearly separated by porosity lines (Fig. 4D). As the water-based suspension spontaneously develops agglomerates with a very broad size distribution (Fig. 3B), its injection into the gas jet during spraying presumably produces an analogously wide agglomerate size distribution. The diethylene glycol-based suspension produces denser coatings, whose individual layers are less clearly recognizable (Fig. 4B).

The HVOF (Fig. 4E, F) coatings have a different microstructure compared to the HVSFS and APS deposited coatings. The coatings consist of unmelted submicron size particles and well flattened lamellae. The unmelted submicron particles (Fig. 4E) originate from the crystallites of the raw powder material. Due to the lower flame temperature of the HVOF, compared to the plasma jet of APS, a significant percentage of the HAp grains did not melt during the HVOF spraying. These porous grains of the raw powder feedstock break up into the earlier mentioned crystallites, and adhere on the coating surface, which are detected by SEM (Fig. 4E). Also a few spherical particles of micron size could be observed on the coating surface, most likely re-solidified crystallites.

The coatings deposited by APS (Fig. 4G, H) have a significant lower amount of unmelted submicron size particles, due to the high temperature of the plasma jet. The APS-HAp coatings predominantly consist of well-flattened lamellae (Fig. 4G). The average diameter of the splats is higher for the APS coatings than for the HVOF ones; although the initial powder feedstock was identical.

X-ray diffraction patterns acquired on the as-deposited coatings leads to the conclusion that the coatings deposited by HVOF are richer in crystalline hydroxyapatite than the ones for APS and HVSFS. Coatings deposited using DEG-based suspensions are richer in crystalline hydroxyapatite than those obtained using water-based suspensions (Fig. 5). All patterns were acquired under the same experimental conditions.

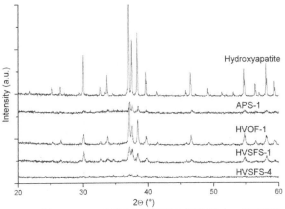

Figure 5. XRD patterns of the as-sprayed APS-1, HVOF-1, HVSFS-1 and HVSFS-4 coatings. The pattern of a pure hydroxyapatite powder is provided for reference. The identified peaks belong to hydroxyapatite (JCPDS card: 09-432).

Surface Roughness and Micro-Indentation tests

For the standard spray processes HVOF and APS the coating roughness is in the same range, due to the identical grain size distribution of the powder feedstock. The HVOF coatings are slightly smoother than the APS coatings due to the higher kinetic energy of the particles, which causes a break up of unmelted grains into submicron particles and flattening of partly melted grains. The high amount of melted particles during APS, resulting in well-flattened splats on the coating surface during deposition, is without much doubt the reason for the smooth surface of the APS-HAp coatings.

Table I. Microstructural and micromechanical properties of the deposited coatings: average ± standard deviation

Coating	Vickers hardness	Elastic modulus [GPa]	R_a [μm]	Deposition efficiency [%]
HVSFS-1	372 ± 33 HV$_{0,1}$	67 ± 5	12.9 ± 1.2	13.0 ± 1.5
HVSFS-2	389 ± 30 HV$_{0,1}$	68 ± 3	12.4 ± 0.6	11.0 ± 1.0
HVSFS-3	283 ± 18 HV$_{0,1}$	51 ± 2	9.5 ± 0.8	26 ± 1.5
HVSFS-4	266 ± 23 HV$_{0,1}$	47 ± 3	9.0 ± 0.5	25 ± 1.5
HVOF-1	277 ± 18 HV$_{0,1}$	60 ± 2	5.5 ± 0.3	42.5 ± 1.5
HVOF-2	373 ± 16 HV$_{0,1}$	80 ± 2	4.2 ± 0.1	57.5 ± 0.5
HVOF-3	356 ± 13 HV$_{0,1}$	75 ± 2	4.4 ± 0.2	55.5 ± 1.5
HVOF-4	384 ± 17 HV$_{0,1}$	72 ± 2	4.7 ± 0.3	80.0 ± 5.0
HVOF-5	357 ± 15 HV$_{0,1}$	73 ± 1	4.4 ± 0.3	55.5 ± 0.5
APS-1	283 ± 13 HV$_{0,1}$	56 ± 1	5.6 ± 0.2	52.5 ± 2.5
APS-2	316 ± 12 HV$_{0,1}$	61 ± 1	5.3 ± 0.2	60.0 ± 1.0
APS-3	323 ± 11 HV$_{0,1}$	62 ± 2	4.9 ± 0.3	62.0 ± 2.0
APS-4	305 ± 10 HV$_{0,1}$	59 ± 1	4.7 ± 0.2	59.0 ± 2.0
APS-5	315 ± 12 HV$_{0,1}$	59 ± 1	5.2 ± 0.2	59.5 ± 1.5

The roughness, summarized in Table I, of the HVSFS-HAp coatings is significantly higher than for HVOF and APS. Fine resolidified spherical droplets and unmelted agglomerates seem to concentrate around asperities of the coating surface. Because of their low inertia, indeed, the fine agglomerates caught in the jet fringes can be deflected by the turbulent flow in front of the substrate; therefore, they travel horizontally along the coating surface and stick preferentially to some prominent asperities. This non-uniform surface characteristic most likely causes the high surface roughness.

Regarding the hardness of the HVSFS-HAp coatings the use of diethylene glycol and water as solvent for the suspension affects the coating hardness (Table I) more than the selection of deposition parameters. As a result of using DEG based suspension the coating hardness is significantly higher than for the water-based ones. The measured Vickers hardness for the HVOF coatings is in the range of the DEG based HVSFS coatings HVSFS-1 and HVSFS-2. Only the Vickers hardness of coating HVOF-1 and APS-1 is as low as the one of the water-based HVSFS-3 and HVSFS-4 coatings. In general it can be said the HVOF-HAp coatings have a higher Vickers hardness than the APS-HAp coatings.

Adhesion Strength

The bond strength values measured using the pull-off method (described in Fig. 1) are displayed in Fig. 6. The bond strength of the deposited HAp coatings is dependent on the spray process and the spray parameters. In literature bond strength values between 5 and 23 MPa for APS [26] and 24 ± 8 MPa for HVOF [12] HAp coatings can be found. Since preparation of the substrate surface has a significant influence on the adhesion strength of the coating composites, all used substrates were treated under same conditions. The highest bond strength of around 25 N/mm^2 was measured for HVSFS- and APS-HAp coatings. The different thermally sprayed layer composites show significant differences in their failure mechanism during the pull-off test. The HVOF and APS coatings adhesively fail at the interface of Ti substrate and coating. The HVSFS-HAp coatings showed a cohesive failure mechanism.

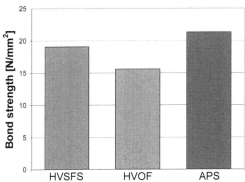

Figure 6. Measured bond strength for the HVSFS, HVOF and APS deposited HAp coatings.

Using DEG instead of water as suspension has a high effect on the bond strength of the HAp coatings, DEG based suspension increases the bond strength compared to water-based one. Also an increased oxygen flow rate improves the performance of the coating composite. The measured values for the HVSFS-HAp coatings are consistent with their microstructure; DEG leads to denser coatings with less interlayer porosity resulting in higher bond strength.

The major effect on the adhesive behavior for HVOF-HAp coatings is the spray distance. Decreasing the spray distance leads to higher bond strength. Fuel flow rate also shows a distinct influence on the coating behavior. Higher fuel flow rate as well as oxygen flow rate leads to higher bond strength.

The bond strength of APS-HAp coatings is also strongly influenced by the process parameters. The main parameter here is the electric arc current, increasing the arc current leads to increased bond strength. Increasing the H2 flow rate as well as reducing the spray distance enhances the bond strength.

CONCLUSIONS

The HVSFS, HVOF and APS techniques were employed to deposit HAp coatings onto titanium plates. The use of different dispersion mediums to prepare the HAp suspensions affects the microstructure and mechanical properties of the resulting HVSFS coatings more than the process parameters. Regarding the influence of rheological behavior of the suspensions on the HVSFS process, it appears that large agglomerates (80 μm peak of the DEG based suspension) do not contribute to the layer composition and instead rebound of the coating surface, resulting in decreased deposition efficiency.

For the HVOF and APS sprayed HAp coatings the deposition efficiency is higher than for the HVSFS process. The HVSFS HAp coatings have the most refined microstructure (Fig. 4A, C) regarding the microstructure of the deposited coatings. XRD measurements indicate a high amount of crystalline hydroxyapatite on the surface of the HVOF and HVSFS DEG sprayed coatings compared to the APS sprayed samples. Verification of the biocompatibility of the HVSFS coatings still has to be

investigated. Simulated body fluid tests could be used as the first step to anticipate the bone bonding ability in vivo of the coatings [27].
Also the main effects for the different processes on the adhesive behaviour were reviewed. The pull-off performance of the HVSFS-HAp coatings strongly depends on the dispersive medium and the oxygen flow rate, for HVOF-HAp coatings it is the spray distance and for the APS-HAp coatings the electric arc current is the most influencing parameter.

Acknowledgements
The authors are particularly grateful to Dr. Baldi and Ms. Lorenzi from Cer.i.col for providing the HAp suspension.

REFERENCES
[1]   S. Kurtz, E. Lau, M. Halpern, K. Ong, Mater Manag Health Care 15(7) (2006) 61-62.
[2]   E. Wintermantel, H. Suk-Woo, Medizintechnik mit biokompatiblen Werkstoffen und Verfahren, ISBN 3-540-41261-1, 3. Aufl. Springer-Verlag, 2002, p 216-228
[3]   C.L. Tisdel, V.M. Goldberg, J.A. Parr, J.S. Bensusan, L.S. Staikoff, S. Stevenson, J. Bone Joint Surg. Am. 76 (1994)159-171.
[4]   J. Dumbleton, M. T. Manley, J Bone Joint Surg Am. 86 (2004) 2526-2540.
[5]   K.A. Khor, P. Cheang and Y. Wang, JOM 49, Issue 2 (1997) 51-57.
[6]   K.A. Khor, H. Li, P. Cheang, Biomaterials 25 (2004) 1177-1186.
[7]   K. Gross, C. Berndt, J. Biomed. Mater. Res. B 39, Issue 4, 580-587.
[8]   R. B. Heimann, Mat.-wiss. u. Werkstofftech. 30 (1999) 775-782.
[9]   S. Dyshlovenko, C. Pierlot, L. Pawlowski, R. Tomaszek, P. Chagnon, Surf. Coat. Technol. 201 (2006) 2054–2060.
[10]   E. Bouyer, F. Gitzhofer, M.I. Boulos, J.Mater. Sci.Lett., 49(2) (1997) 58-62
[11]   C. Renghini, E. Girardin, A.S. Fomin, A. Manescu, A. Sabbioni, S.M. Barinov, V.S. Komlev, G. Albertini, F. Fiori, Mater. Sci. Eng. B 152 (2008) 86-90.
[12]   R.S. Lima, K.A. Khor, H. Li, P. Cheang, B.R. Marple, Mat. Sci. Eng. A 396 (2005) 181–187
[13]   C. Monterrubio-Badillo, H. Ageorges, T. Chartier, J.F. Coudert, P. Fauchais, Surf. Coat. Technol. 200 (2006) 3743–3756.
[14]   F-L. Toma, G. Bertrand, D. Klein, C. Coddet, C. Meunier, J. Therm. Spray Technol. 15 (2006) 587–592.
[15]   F-L. Toma, G. Bertrand, S. Begin, C. Meunier, O. Barres, D. Klein, C. Coddet, Appl. Catal. B 68 (2006) 74–84.
[16]   A. Killinger, M. Kuhn, R. Gadow, Surf. Coat. Technol. 201 (2006) 1922–1929.
[17]   X.Q. Ma, J. Roth, D.W. Gandy, G.J. Frederick, J. Therm. Spray Technol. 15 (2006) 670–675.
[18]   P. Fauchais, R. Etchart-Salas, V. Rat, J.F. Coudert, N. Caron, K. Wittmann-Ténèze, J. Therm. Spray Technol. 17 (2008) 31-59.
[19]   R. Rampon, F.-L. Toma, G. Bertrand, C. Coddet, J. Therm. Spray Technol. 15 (2006) 682–688.
[20]   R. Gadow, A. Killinger, M. Kuhn, D. López, Hochgeschwindigkeitssuspensionsflammspritzen, Deutsche Patentanmeldung Nr. DE 10 2005 038 453 A1
[21]   J. Rauch, N. Stiegler, A. Killinger, R. Gadow, in: B.R. Marple, M.M. Hyland, Y.-C. Lau, C.-J. Li, R.S. Lima, G. Montavon (Eds.), Thermal Spray 2009: Expanding Thermal Spray Performance to New Markets and Applications - Proceedings of the International Thermal Spray Conference, ASM International, Materials Park, OH, USA, 2009, pp 150-155.
[22]   A. Killinger, M. Kuhn and R. Gadow, Surf. Coat. Technol. 201 (2006) 1922-1929
[23]   R. Lietzow, Herstellung von Nanosuspensionen mittels Entspannung überkritischer Fluide, Doctoral Thesis (2006), Universität Karlsruhe, ISBN 978-3-86727-080-9, Cuvillier-Verlag

[24]   C. Synowietz, K. Schäfer: Chemiker-Kalender, 3. Edition, Synowietz C., Schäfer K. (ed.),
        Springer-Verlag, 1984, ISBN 3-540-12652-x
[25]   L. Pawlowski, Surf. Coat. Technol. 202 (2008) 4318-4328
[26]   S.W.K. Kweh, K.A. Khor, P. Cheang, Biomaterials 21 (2000) 1223 -1234
[27]   M. Bohner, J. Lemaitre, Biomaterials 30 (2009) 2175–2179

Coatings to Resist Wear, Erosion, and Tribological Loadings

CERAMIC / METAL - POLYMER MULTILAYERED COATINGS FOR TRIBOLOGICAL APPLICATIONS UNDER DRY SLIDING CONDITIONS

A. Rempp*[a], M. Widmann[b], P. Guth[b], A. Killinger[a] and R. Gadow [a]
[a] IFKB, University of Stuttgart, Stuttgart
[b] WITTENSTEIN bastian GmbH, Fellbach

ABSTRACT

The combination of thermally sprayed hard metal or oxide ceramic coatings with a polymer based top coat leads to bi or multilayered coating systems with tailored functionalities concerning wear resistance, friction, adhesion and wetting behaviour or specific electrical properties. This type of coatings is successfully inserted in numerous industrial applications. The basic concept is to combine the wear resistance and friction behaviour of the basic hard coating with the tribological or chemical abilities of the polymer top coat suitable for the respective application. This paper gives an overview of different types of recently developed multilayer coatings and their application in power transmission under dry sliding conditions. To evaluate the capability of these coatings, steel substrates were coated with oxide ceramics, metal alloys and hard metals by high velocity flame spraying (HVOF). After a grinding process several types of sliding lacquers are applied by air spraying on hardened steel and coated specimens. Different sliding lacquers based on polyamide-imide filled with solid lubricants e.g. MoS_2, PTFE or graphite were tested under high loads and sliding conditions. Wear resistance and friction coefficients of combined coatings were determined under dry sliding conditions using a twin disc test-bed.

1 INTRODUCTION

Nowadays a big issue in industrial applications is the sustainable handling of material and energy resources. Due to the demanding environmental restrictions, mechanical components and tools are facing higher requirements. For numerous industrial applications the realization of several, sometimes opposing, physical or chemical surface properties are required. For tribological applications the reduction and replacement of lubricants, e.g. mineral oils and grease, is an important issue. Reduction of service and maintenance costs is another motivation for the process and product development.

Usually tribological systems are described by external parameters such as contact load, contact area, material of the friction partners, surface textures, lubricant used and the environmental parameters (humidity, temperature, etc.). The coefficient of friction and the wear are the main parameters used to describe the performance of a tribological system. When the lubricant is removed the requirements to a tribological system change and some of the important functionalities of the lubricant like cooling and separation of the contact surfaces have to be transferred to other components. Often the substrate materials, especially light metal alloys, cannot meet the structural requirements under dry sliding conditions. Even the smoothest technical surfaces are rough on the micrometric scale and so the apparent contact area is much larger than the true contact area. The sliding partners only touch at the high spots of their surface topography and due to the small bearing area the real surface load is sometimes higher than the structural resistance of the substrate. The friction behavior of dry sliding systems is depending on the substrate or coating materials and is mainly characterized by adhesion and abrasion. So in a metal-metal contact shearing welded junctions are formed between points of contact and ploughing out surfaces of the softer material by the harder material. For hard metals the ploughing part is relatively small and so the friction behavior is due to the shearing of welded junctions [1]. To reduce the load of a dry sliding system, materials have to be applied which come up with very smooth surfaces, high hardness and low adhesion tendency to prevent abrasive and adhesive wear.

A promising way to realize those systems is to cover the substrate material with coatings. Its use opens up the possibility to a material design in which the specific properties are located where they are most needed. Therefore, a lot of research was done on the development of self-lubricating and wear resistant coatings. Thin coatings based on amorphous carbon have become widely accepted for reducing friction and wear in engineering applications [1]. One of the most promising coating type is diamond-like carbon (DLC). The term DLC covers a wide range of different carbon-based coatings. The properties of these coatings vary due to the coating composition. Depending on the proportion of the carbon's tetragonal phase, the coatings can be either very hard and wear resistant or polymer-like with lower hardness and lower friction coefficients. The influence of the coating hardness and surface roughness on tribological behavior of DLC coatings have been investigated in [2] and [3]. These coatings can also include amorphous materials with a content of up to approximately 40 at.-% hydrogen (a-C: H) and materials that contain less than 1 at.-% hydrogen (a-C) [4]. Another type of thin coatings is based on solid lubricants, e.g. MoS_2 or graphite. It allows contacting surfaces to run against each other with reduced friction and wear. A drawback of thin coatings is the limitation of the coating thickness, which limits the permissible wear volume. Two carbon-based solid lubricant coatings have been investigated in [5]. These coatings are able to bear high contact pressures and offer excellent tribological properties. Furthermore they protect the counterpart by providing a low friction transfer film [6,7].

Another possibility to improve the tribological properties of materials beside PVD-coatings is to apply thermally sprayed coatings [8,9,10]. These coatings consist of metal alloys, ceramic and cermets - especially those based on oxides show high wear resistance and some of the coating materials have self-lubrication abilities.

In the recent year, at the IFKB a lot of research has been done on the issue of dry sliding systems. To realize a dry sliding tribological system different coating composites have been investigated. The coating composites were based on the principle of Ceramic / Polymer composite coatings. The fundamentals of these coatings are described in [11] and [12]. The basic idea of these composite coatings is to obtain tailored surface properties by combining different coating techniques and materials.

In this paper, a new approach to realize a dry sliding system according to the principle of composite coatings is investigated. The evaluated tribological system is a combination of different coating materials and coating techniques. The composition of this system is shown in Figure 1.

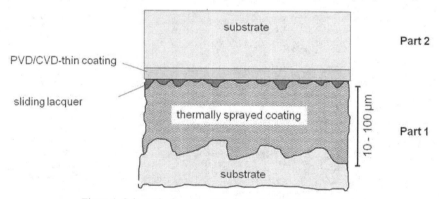

Figure 1. Schematic drawing of the evaluated tribological system

The tribological system is based on a combined ceramic or metal polymer coating system on part 1 and a diamond-like carbon coating on part 2., The surface properties can be adjusted to the specific requirements of the application by using different coating techniques. Steel, iron based alloys and light metals can also be used for substrate materials, as well. A lubricant lacquer deposited on top of the thermal sprayed coating covers the peaks and cavities of the sprayed surface to reduce the friction coefficient. Since the lubricant lacquer is applied in very thin layers, the properties of the hard phase coating as well as the properties of the lubricant lacquer can take effect on the contact surface. Lubricant lacquers are especially used in industrial applications where grease and lubricants fail to function, like in vacuum, at extremely low temperatures, at high temperatures and low pressures, in the presence of solvents, etc. [11]. Another important field of application is small relative motion under high surface pressure, which causes high static friction.

2 EXPERIMENTAL WORK

Cylindrical steel specimens are used to evaluate the properties of the tribological system. The specimens have a diameter of 41 mm and a width of 15 mm. The substrate material is case-hardened steel. The test program is separated in two test steps. For the first step test samples are hardened and a defined surface quality and shape is adjusted by grinding. Based on the results of former investigations [11] [12] [13] the ground surface is textured by laser. In this study, four different textured surfaces (Table I) are compared to a none textured one. The adjusted surface topography differs in amount and shape of the laser cavities. Figure 2 shows an example of a textured surface and a 3D image of a laser dimple.

Table I. Overview of the laser texturing

Abbreviation	Amount of cavities per track	Volume of removed material
LS-I	5000	low
LS-II	5000	high
LS-III	2000	low
LS-IV	2000	high

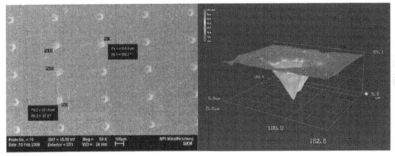

Figure 2. SEM image of a textured surface and 3D image of a laser dimple

As a next step, the specimens are coated with different sliding lacquers by air spraying. The composition of the lacquers is varied concerning binder content, solid lubricant and particle size of the lubricant. The polymer is based on polyamide-imid, all systems described in this paper are based on the same binder system. Table II shows the chosen solid lubricants and its average particle size. The

Ceramic/Metal-Polymer Multilayered Coatings for Tribological Applications

binder content is varied in three steps. The systems with a low proportion of binder are described by the abbreviation B1, the systems with medium proportion B2 and with high proportion B3. The sliding lacquers contain 10 and 70 wt.-% binder.

Table II. Solid Lubricants used in sliding lacquer systems

Abbreviation	Material	Particle Size (D50)
MoS2 – f	MoS2	8 μm
MoS2 - c	MoS2	20 μm
BN – f	BN	0,5 μm
BN – c	BN	5 μm
PTFE – f	PTFE	4 μm

In the second test stage, the influence of the thermally sprayed material on the evaluated composite coating system is investigated. Therefore the best combinations of sliding lacquer and surface pre-treatment is chosen and tested on different thermal spray coatings. The steel substrates were coated in High Velocity Oxygen Fuel Thermal Spray process (HVOF) and High Velocity Liquid Fuel Thermal Spray process (HVLF) with different coating materials based on metal alloys, ceramics and cermets.
Table III shows data of the applied spray powders.

Table III. Thermal Spray Powders

Abbreviation	Material	Composition	Particle Size	Coating Thickness (after grinding)	Coating Hardness (HV0.1)
Cr2O3	Chromium Oxide	100	-25 + 5 μm	~ 25 μm	1508.82
WC/CoCr(FC)	Cermet (Fine Carbides)	86/10/4	-45 + 15 μm	~ 25 μm	1392.93
Mo	Molybdenum	99/1(O_2)	-45 + 20 μm	~ 25 μm	770.54
Cu-Ni-In	Copper-Nickel-Indium	60/35/5	-45 + 11 μm	~ 25 μm	271.09

The tribological tests were performed on a two disc tester. A radial load is applied and the specimens revolve, while its axes remain parallel. The running tracks are hence on the circumferential surface of each disc. Different rotating speeds of the discs lead to a movement with rolling and a sliding.

42 · Advanced Ceramic Coatings and Materials for Extreme Environments

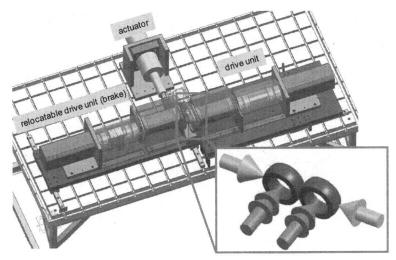

Figure 3. Test setup on the two disc tester with indicated radial force (single arrowhead) and drive torque (double arrowhead)

Slip is adjusted to 20%. A test cycle runs until the friction coefficient exceeds 0,4 for at least 2 seconds. Cycles-to-failure and mean friction coefficient are the output used for comparisons of the tested systems. For all tests one specimen was coated with a polymer hard phase composite coating and the counterpart was coated with a commercial available DLC coating (DLC-Star©, Balzers Oerlicon). The DLC-Star coating (DLC) is well-known for its good tribological properties, especially for their low friction coefficient under dry sliding conditions.

3 RESULTS AND DISCUSSION

In Figure 4 a cross section of a sliding lacquer before and after running-in is shown. There is good contact between the polymer and the steel substrate. During running-in, excess polymer and solid lubricant is removed and solified until a stable lubricating film is formed.

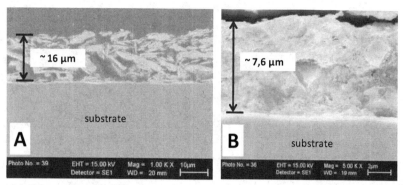

Figure 4. Cross section of a sliding lacquer coating MoS₂ flakes before (A) and after (B) running-in

In Figure 5 a cross section of a metal/lubricant lacquer composite layer before and after 3500 revolutions is shown. The contact between the polymer and the metal layer is still good and the applied polymer considerably follows the surface profile. The lamellar structure typical for thermally sprayed coatings is clearly visible. During operation the excess polymer is removed so that it only fills the pores and structures on the thermal spray coating. This composite now forms the contact area.

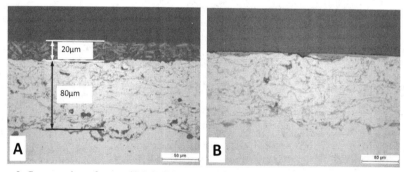

Figure 5. Cross section of a metal/MoS₂ lubricant lacquer composite layer before (A) and after (B) testing

As shown in Figure 5 B there is still a thin film of lubricant lacquer present at the sliding interface which causes reduction of the shear forces and prevents opposing asperities from coming into direct contact. Easy shearing at contact interfaces results in low friction, whereas fewer asperity interactions means reduced wear. Even though there is still sliding lacquer in the pits of the thermal spray coating, the life-time of the composite coating is lower than 10,000 revolutions. Due to the grinding of the thermal spray coating the roughness of the surface is too low and so there are only small structures available to use as reservoir for the lubricant lacquer on the surface.

Tribological properties of lubricant lacquers

Based on first test results as shown in Figure 4 and Figure 5, the first test stage was planned. This investigation was focused on the influence of the surface pre-treatment on the tribological behavior of sliding lacquers. In addition the influence of the binder content and particle size of the lubricants was evaluated. Because of the large number of parameters the test planning has been worked out according to DOE-methods to reduce the number of tests. In Figure 6 the correlations between the different parameters, the friction coefficient and the life-time are shown.

a

b

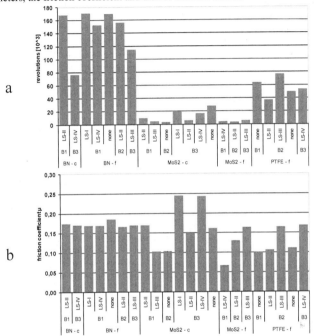

Figure 6. Correlation between solid lubricant, binder content, surface pre-treatment and tribological properties of lubricant lacquer coatings; (a) life-time of sliding lacquers vs. DLC, (b) friction coefficient of sliding Lacquers vs. DLC

All test samples were grinded to a surface roughness of R_a=0.8 μm before texturing. After the laser treatment the samples with the structures LS II and LS IV were grit blasted to remove the process-related bulging. Subsequently the specimens were coated with different sliding lacquers. The thickness of the sliding lacquers is in the range of 10 to 25 μm.

A look at the results of the lacquer samples shows that there are significant differences in the correlations between the binder content, the surface treatment and the tribological properties of the sliding lacquer depending on the solid lubricant. The dry sliding and lubrication ability of Boron nitride (BN) seems to be almost unaffected by the surface topology and the composition of the lacquer. BN reaches very long life-time values although the friction coefficient is quite high. Because of the long life-time which already is higher than the values of the reference system, BN is a promising candidate for a combined coating system. In comparison to BN the other solid lubricants show strong dependencies on the evaluated parameters and only poor tribological properties. Although MoS_2 is well-known as solid lubricant for high pressure applications, in case of a combined high pressing and high sliding velocity, MoS_2 seems to reach structural limits which leads to the low life-time values. Even if the life-time values of PTFE are better compared to the results of MoS_2, PTFE is not able to reach the good results of BN.

In Figure 7 the measurement result of the DLC-coating is shown in comparison to a BN-coating. The life-time value of BN is more than 30,000 revolutions above the value of the DLC. The DLC shows its typical low friction coefficient over long period but the rising of the friction coefficient is already starting at around 100,000 revolutions. At this point it seem that the DLC coating has been removed from the surface by abrasive wear and now the wear mechanism changes from abrasion to adhesion. A closer look to the curve of BN shows that the average friction coefficient has a much lower variance, its value stays in the range of 0.15 to 0.2 until failure occurs.

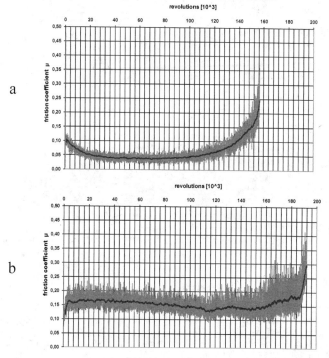

a

b

Figure 7. Measurement recording of a DLC (a) and a BN(LS I;B1)-lacquer coating (b)

Based on the results of the first test stage the best sliding lacquer compositions have been chosen for further investigation steps. To evaluate the influence of the substrate material there was also a sliding lacquer with PTFE chosen for the next test step. Both sliding lacquers are applied in combination with the LS I laser texture. The PTFE lacquer composition is B2 with fine particles and the BN lacquer composition is B1 also with fine particles.

Tribological properties of combined coating systems

For the second test step four different thermal spray systems were applied. Because of the high hardness and wear resistance of the thermally sprayed coatings, all specimens were grinded to a surface roughness of $R_a = 0.25^{+/-0.05}$ µm.
In Table IV a cross section of each coating system before and after testing is shown. The lacquer coating thickness was adjusted to 10 to 20 µm depending on the lacquer system.

Table IV. Cross sections of Ceramic / Metal Polymer composite coatings

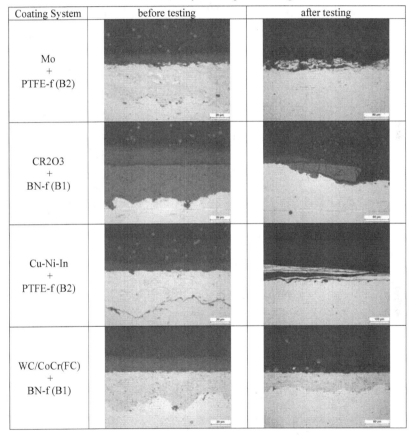

Coating System	before testing	after testing
Mo + PTFE-f (B2)		
CR2O3 + BN-f (B1)		
Cu-Ni-In + PTFE-f (B2)		
WC/CoCr(FC) + BN-f (B1)		

As it can be seen most of the composite coating systems fail due to the extreme abrasive wear. Depending on the substrate material the damage mechanism seem to differ. The cross sections of Mo and CR2O3 show that the coatings have been totally broken up and removed. Also the cross section of Cu-Ni-In shows delamination but compared to the failure mode of MO and CR2O3 the abrasive wear takes place layer by layer. In Figure 8 the average life-time and the friction coefficients of the evaluated coating systems are shown.

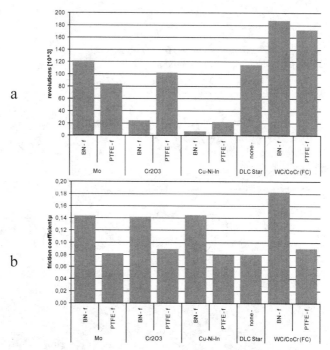

Figure 8. Correlation between substrate material, sliding lacquer and tribological properties of combined coatings, (a) life-time of composite coatings vs. DLC, (b) friction coefficient of composite coatings vs. DLC

A part from Cu-Ni-In which showed poor results independent of the applied sliding lacquer poor results, some combined coating systems have reached better results compared to the results from the sliding lacquer tests. The poor results of Cu-Ni-In are owing to the layer by layer removal which leads to a delamination of large coating areas and so to a fast increase of the friction coefficient. The results of BN based systems show large variations. Depending on the substrate, most of the BN composites did not reach the values of the pure polymer layers. The curve shape of a BN/WC/CoCr(FC) composite coating (Figure 9) is comparable to the graph shown in Figure 7 b. The graph shows that the friction coefficient is mainly influenced by the solid lubricant and almost unaffected by the substrate.

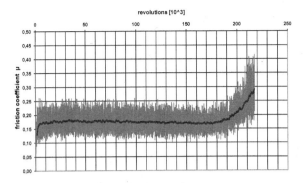

Figure 9. Measuring record of a BN/WC/CoCr(FC) composite coating

There is only a slightly different behavior at the point of failure. While the sliding lacquer on the steel substrate fails sudden and unpredicted, the failure of the composite coating is signalized by a slow rise of the friction coefficient. Even if the life-time of the composite coating is above 210,000 revolutions which is an increase of over 10 % the results did not meet the expectations. To understand the working principle of BN further investigations have to be done. The PTFE based composite coatings show unlike the BN systems a significant improvement of their tribological behavior depending on the substrate material. Apart from Cu-Ni-In composites all composite coatings have reached life-time values above 80,000 revolutions. Surprisingly the PTFE/WC/CoCr(FC) composite was also able to exceed the life-time of reference system. In Figure 10 a measurement plot of the PTFE/WC/CoCr(FC) composite is shown. The curve shape is quite different compared to the graph of BN/WC/CoCr(FC). There is a rise of the friction coefficient at the beginning is inexplicable.

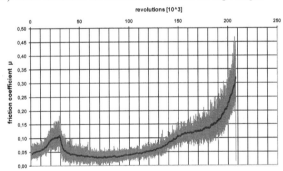

Figure 10. Measurement plot of a PTFE/WC/CoCr(FC) composite coating

Between 40,000 and 140,000 revolutions a stable status is reached with a friction coefficient which is comparable to the reference system. Due to the abrasive wear the value increases slowly to a higher level and at around 180,000 revolutions the coating shows incipient failure.

4 CONCLUSION

It has been demonstrated that combined polymer hard phase coating systems are a powerful concept for dry sliding applications where high wear resistance, creep resistance and compressive strength as well as low friction coefficients are required. Under the specific experimental conditions used in this study, the BN and PTFE based layers in combination with the applied cermet coating showed better performance than the DLC coatings. Further studies have to be done using improved lacquer compositions as well as other hard materials.

Further modification and variation of the thermally sprayed layer as well as the applied polymer will lead to coating systems which may be used in numerous other fields within mechanical engineering, offshore applications and the consumer goods industry.

5 REFERENCES

[1] Francis J. Clauss; Solid Lubricants and Self-Lubricating Solids, Academic Press (1972)

[2] J. Brand, C. Brand, J. Gäbler, Tribologie + Schmierungstechnik, 50, 18 (2003).

[3] Per Lindholm, Stefan Björklund, Fredrik Svahn; Wear 261, 107–111 (2006)

[4] R. Gadow, D. Scherer, Surface and Coatings Technology 151 –152, 471–477 (2002)

[5] S.K. Field, M. Jarratt, D.G. Teer; Tribology International 37, 949–956 (2004)

[6] Yang S, Camino D, Jones AHS, Teer DG. Deposition and tribological behaviour of sputtered carbon hard coatings. Surface and Coatings Technology, 124:110 (2000)

[7] Jarratt M, Stallard J, Renevier NM, Teer DG. An improved diamond-like carbon coating with exceptional wear properties. Diamond and Related Materials, 12(3–7), 1003–7 (2003)

[8] C. Bartuli, T. Valente, F. Casadei, M. Tului; IMechE, Vol. 221, 175-185, Part L: J. Materials: Design and Applications (2007)

[9] M. Woydt and U. Effner; Machining Science and Technology, 1(2), 275-287 (1997)

[10] L.-M. Berger, C.C. Stahr, S. Saaro, S. Thiele, M. Woydt, N. Kelling; Wear 267, 954–964 (2009)

[11] R. Gadow, D. Scherer, Surface and Coatings Technology 151 –152, 471–477 (2002)

[12] D. Scherer, R. Gadow, A. Killinger; United Thermal Spray Conference, Düsseldorf; Conference Proceedings; E. Lugscheider, P.A. Kammer (ed.), pp. 664-669; ISBN 3-87155-653-X (1999)

[13] P. Guth; Tagungsband FTK 2010, 519-535 (2010)

APPLICATION OF HVOF FOR HIGH PERFORMANCE CYLINDER LINER COATINGS

A. Manzat, A. Killinger, R. Gadow
Institute for Manufacturing Technologies of Ceramic Components
and Composites, University of Stuttgart
Stuttgart, Germany

ABSTRACT

Rising demands for low emissions automotive engines require a significant decrease in fuel consumption and internal friction. One approach to achieve this goal is the development of highly loaded and downsized internal combustion engines. This accentuates the need for highly wear resistant materials. In addition, the deployment of bio-fuels requires corrosion resistant materials.

New technologies for the processing and application of advanced materials for improved tribofunctional coatings for internal combustion engines are described. These coatings contribute to reduce the frictional losses and improve the wear and corrosion resistance much needed. HVOF (High Velocity Oxy Fuel) sprayed metallic (Fe, FeCrMo) and cermetic (Cr_3C_2/NiCr, n-WC/Co) and HVSFS (High Velocity Suspension Flame Spray) sprayed ceramic (TiO_2/TiC) coatings will be presented. The HVSFS process represents a novel method for the direct processing of nanoscale materials which opens new application fields even for established materials since the nanoscale structure can show improved properties compared to the respective standard coatings. Selected test results of the novel coatings compared to state-of-the-art materials will also be presented.

INTRODUCTION

One major approach for low emissions and reduced fuel consumption is the deployment of complex and expensive AlSi-alloys showing advanced structural stability, but on the other hand increase the casting efforts.

An approach to solving these problems is the application of thermally sprayed coatings on inferior but easy to process Al-alloy for engine blocks and thereby increasing their aptitude for a use as lightweight crankcases with improved tribological properties. Most of the approaches so far include the use of specially adapted, small scale torches, e.g. internal rotating spray torches based on Atmospheric Plasma Spraying (APS), Wire Arc or Plasma Transfer Wire Arc (PTWA) techniques [1, 2].

In this research a standard HVOF torch has been used, which facilitates the application of high quality coatings and also overcomes the limitations in feed stock choice. Contrary to internal rotating torches, the HVOF torch is operated outside the cylinder bore with the cylinder liner rotating [3].

EXPERIMENTAL

The coatings were deposited on aluminum cylinder liners and inline-4-cylinder engines representing the typical size of passenger car engines (82 mm bore, 32,500 mm^2 coated surface) and gray cast iron liners for large commercial vehicle engines (131 mm bore, 82300 mm^2 coated surface). Deployed coating materials are shown in Table I.

All liners have been grit-blasted prior to the coating process. The coatings have been applied by means of a HVOF (High Velocity Oxy Fuel) process using the GTV "Top Gun G" torch with propane as combustion fuel. TiO_2/TiC-coatings have been handled in a suspension due to the small particle size and therefore been applied using the High Velocity Suspension Flame Spraying (HVSFS) process. A modified GTV "Top Gun G" torch has been used for this application [4].

Table I. Deployed coating materials

Brand Name	Manufacturer	Composition	Particle Size (d,50)
Nano3	Durum	Fe-alloy	25 μm
Diamalloy 1008	Sulzer-Metco	FeCrMo	36 μm
Infralloy 7412	Inframat	WC/Co	31 μm
P25	Evonik	TiO_2	0.02 μm
STD 120	H.C. Starck	TiC	2 μm

All internal coatings have been deposited with the torch being manipulated by a computer-controlled 6-axis robot. The trajectory of the torch can be described as an ellipsoidal movement to account for the geometrical constraints imposed by the process of internal coating with an externally mounted torch and the need of a constant spraying distance and constant step size on the coated surface in order to realize homogeneous properties of the coatings.

Since the deposition efficiency is depending on the spraying angle and the thermal conditions, elaborate adaptations of the torch kinematics are necessary for being able to achieve a constant coating thickness. The main adjustments made to control thickness are adjustments to the speed of the torch movement along the predetermined trajectory as the speed of passing over a surface also influences the resulting coating thickness. This generally results in a regular coating thickness with deviations as low as ± 6 %, meaning below ± 20 μm for a 300 μm coating, depending on the deployed coating material.

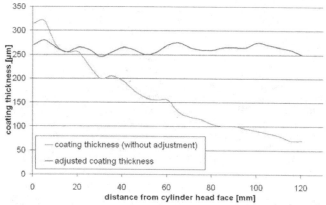

Figure 1. Exemplary coating thickness gradient before and after speed adjustment

This adaptation also resulted in significantly lower processing times as there was no need for additional allowance to account for deviations in the coating and also reduces the costs for mechanical treatments, e.g. lower stock removal during honing.

Another geometrical requirement affected the upper ending of the coating. Due to the limited resistance against possible lateral stresses applied by the cylinder head, the coating has to run out at least one mm under the cylinder head face to ensure a non-contacting operation during the engine run. This issue has been solved by using a specially developed masking, which led to a gradually decreasing coating thickness with a smooth edge on the upper coating end.

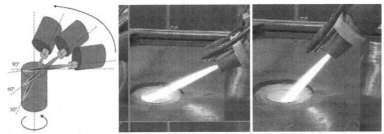

Figure 2. Schematic movement of the torch during the coating of a cylinder liner (left); actual torch position for coating thr top dead center area (middle) and bottom dead center area (right)

The coating has been applied by means of several torch passes with a deposition of up to 74 μm per pass depending on coating powder, resulting in an average coating thickness of 300 μm respectively 200 μm for TiO_2/TiC. Despite the disadvantageous spraying angles, acceptable overall deposition efficiency could be achieved, especially for the smaller aluminum liners, the deposition efficiency reached more than 55 %. All trajectories and speed adaptation applied during liner coating can be transferred to engine block coating without any modification leading to the same coating thickness shapes. After being coated, the liners have been honed with the same crossing groove structure like the state-of-the-art gray cast and Al-alloy liners, except that the grooves could be formed significantly shallower. The pores on the surface form micro cavities which improve the oil retention abilities superseding the deep honing grooves. This results in a smooth surface structure with small pores which are typical for thermally sprayed coatings [5].

The HVOF and HVSFS processes require a standoff distance of around 200 to 250 mm and 100 to 150 mm respectively. These standoff distance ranges enable the coating of cylinders having a diameter from 30 up to 250 mm and with an aspect ratio of up to 1.5. The presented coating setup therefore permits the coating of very small downsized engines as well as liners for commercial vehicle engines with the same equipment achieving the same coating properties.

MATERIAL ANALYSIS AND CHARACTERIZATION
The material screening was focused on the development of high-efficiency coating systems for tribological applications with regard to low friction and wear coefficients and high corrosion resistance as well as enhanced operational performance.

The micro hardness has been determined with a micro-indenter from Fischer, calculating a HV0.1 micro hardness according to DIN 50359 / ISO 14577 tested on polished cross sections of the samples. The porosity of the applied coatings has been determined by digital image analysis of cross section micrographs.

Table II. Porosity and micro hardness

	Fe-alloy	FeCrMo	$Cr_3C_2/NiCr$	NiCrBSi	TiO_2/TiC	WC/Co
av. porosity [%]	1.1	2.9	1.2	1.2	1.4	0.1
HV0.1	810	710	1000	775	800	1330

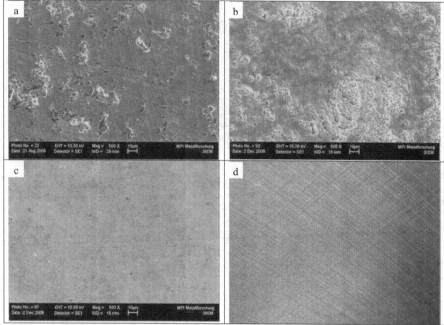

Figure 3. Exemplary SEM-images of honed surfaces of Cr_3C_2/NiCr (a), TiO_2/TiC (b), WC/Co (c) compared to conventionally honed gray cast iron (d)

Table III. Coated surface finish characteristics after honing

	Fe-alloy	FeCrMo	Cr_3C_2/NiCr	TiO_2/TiC	WC/Co
Ra	0.20	0.13	0.16	0.21	0.02
Rz	4.99	3.30	3.58	3.19	0.24
Rpk	0.05	0.05	0.06	0.17	0.02
Rvk	1.06	0.71	0.78	0.82	0.03

TRIBOLOGICAL ANALYSIS

Ball-on-Disk Tests

Ball-on-disk tests have been performed with both coated and non-coated liners as a comparison to state-of-the-art gray cast and Al-alloy liners. All tests have been conducted on honed surfaces and under ambient conditions (22 °C).

The first ball-on-disk test has been conducted under non-lubricated conditions with an alumina ball as counterpart to accentuates the wear and load on the tested sample in order to have a time lapse effect for an easier calculation of the wear coefficient K_V. This test doesn't represent the frictional conditions occurring during engine operations but it has been chosen for a comparative evaluation of the wear resistance of the coatings. The friction tests have been performed during 100,000 cycles at 10 N load with oscillating motion of 5 mm displacement at a speed of 70 mm/s. After the friction tests the wear volume has been determined and the wear coefficient has been calculated according to equation 1 [6], where K_V represents the volumetric wear coefficient [mm³/Nm], V_e the wear volume [mm³], F_N the normal load force [N] and s the sliding distance [m].

Table IV. Calculated volumetric friction coefficients K_V

	Fe-alloy	FeCrMo	Cr₃C₂/NiCr	TiO₂/TiC	WC/Co	GCI	AISI
K_V [10^{-6} mm³/N·m]	6.0	1.0	0.2	15.0	0.1	1.6	23.7

$$K_V = \frac{V_e}{F_N \cdot s}$$

(1)

In addition a fretting test has been performed at increasing load starting at 10 N and of 55 N. The speed was 70 mmps with an oscillating motion of 5 mm displacement. A 100Cr6 ball (a material that piston rings often are made of) with 5 mm diameter has been used as counterpart for the samples because of its higher fretting tendency leading to shorter failure times. For these tests, a poor lubrication environment has been simulated. The lubricant (motor oil 15W-40) has been applied once at the beginning of the test to the 100Cr6 ball.

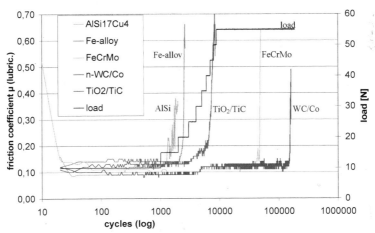

Figure 4. Friction coefficient progression (fretting test); coatings on aluminum liners

The gray cast iron sample showed an increasing friction coefficient after approximately 100 cycles, which is the indicator for incipient failure. The AlSi17Cu4 sample resisted for 1,250 cycles before showing signs of failure. These two samples are state-of-the-art and are being used in production passenger car engines. The FeCrMo-coated sample showed a constant friction coefficient with a light increase until failure. The TiO₂/TiC coated sample showed a constant but rather high friction coefficient before fretting occurred. Outstanding performance has been achieved with the WC/Co coating which lasted for 150,000 cycles having a very low friction coefficient.

For the gray cast liners a similar result has been achieved from the fretting tests. The non-coated gray cast sample showed an instantly increasing friction coefficient resulting in failure. The Fe-alloy sample lasted for 4,300 cycles before incipient failure. The $Cr_3C_2/NiCr$ coating showed a constant coefficient for 5,000 cycles with a slight increase until failure commenced. A similar behavior could be observed for the TiO_2/TiC coating except for the higher friction coefficient throughout the fretting. The noticeable difference between the TiO_2/TiC coatings on Al-Alloy liners and gray cast iron liners result from the different liner material hardness and the different coating conditions due to the varying geometrical properties.

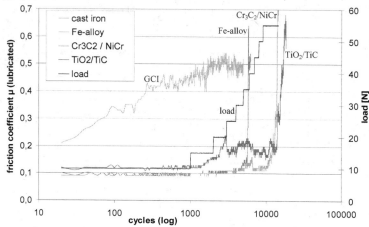

Figure 5. Friction coefficient progression (fretting test); coatings on gray cast iron liners

Piston Ring Tests

In addition to the abstract testing method of ball-on-disk tribometry, piston ring-on-cylinder liner segment testing has been carried out to have a more representative tribological environment similar to the conditions during an engine run. These tests have therefore been performed under controlled lubricated conditions, where oil temperature and piston ring loads have been varied to simulate different states of engine load and speed. The temperature and load values ranged from around 90 to 110 °C and 150 to 450 N respectively.

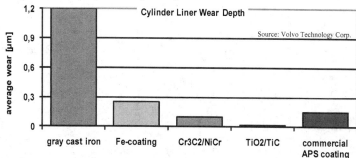

Figure 6: Average cylinder liner wear after piston ring on cylinder liner testing

After the tests, the wear on the coated cylinder liner segments has been identified and compared to state of the art non-coated gray cast iron liners by measuring the radial material loss occurred during the testing. The piston ring testing affirms the results of the ball-on-disk tests, showing a significant improvement in wear resistance with thermal spray coatings.

Engine Test

In addition to the laboratory testing, full engine tests has been conducted to prove the endurance and the functionality of the coatings under real engine run conditions. Drag torque measurements have been done on an inline-4-cylinder motorcycle engine (600 cm³ displacement) to determine the friction reduction when using coated cylinders vs. series production standard Nikasil-coated engine (see figure 8).

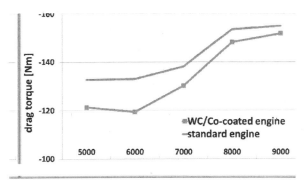

Figure 7: Drag torque of the WC/Co-coated engine vs. Nikasil-coated standard engine

Another test consisted in a continuous engine run for almost ten days comprised of different engine load and speed states. The evaluation showed that the coatings have the potential of enduring a complete engine life expectation, but also showed that further adaptations have to be made to the tribological system to fully take advantage of the benefits of the coatings. On top, the engine run demonstrated that the coatings offer the potential to reduce the oil consumption of internal combustion engines validating the theory.

CONCLUSION

The tribological behavior of surfaces in internal combustion engines can be improved by metallic and ceramic protective coatings. Tribological investigations show low friction coefficients and an excellent wear resistance, especially under deficient lubrication conditions. HVSFS was successfully used to apply fine structured, tribologically functional coatings on aluminum and gray cast cylinder liners. This new approach offers the possibility of deploying new coating materials and compositions [7].

The only structures on the gray-cast and Al-alloy samples are the deep honing grooves, whereas during the honing process the pore structure of the spray coatings become exposed and result in locally confined non-adjunctive micro cavities on the surface. These micro cavities aid in holding oil and therefore form micro pressure chambers supporting the hydrodynamic state of friction reducing wear and the friction coefficient. This especially helps during friction with low relative motion speeds between liner surface and piston ring occurring near the reverse points. In addition the micro pressure chambers reduce the tendency of fretting in deficient lubrication states of the tribological system, e.g. spirited driving with low oil level or under rough off-road conditions. In addition, a significant improvement in the durability of the engines can be expected according to the very low wear coefficients of the applied coatings. By deployment of the specially adjusted flame spray torch kinematics the layer deposition can effectively be controlled, which results in lower processing times during coating and reduces after-treatment efforts. In addition, the coatings can reduce the oil consumption of internal combustion engines and therefore reduce polluting emissions resulting from burned lubricant in the cylinder.

REFERENCES

[1] G. Barbezat, S. Keller and G. Wuest, Internal plasma spray process for cylinder bores in automotive industry, Proceedings of the 15th International Thermal Spray Conference, Nice (1998)

[2] K. Bobzin, F. Ernst, J. Zwick, T. Schlaefer, D. Cook, K. Nassenstein, A. Schwenk, F. Schreiber, T. Wenz, G. Flores and M. Hahn, Coating Bores of Light Metal Engine Blocks with a Nanocomposite Material using the Plasma Transferred Wire Arc, Journal of Thermal Spray Technology (2008)

[3] M. Buchmann, R. Gadow, A. Killinger and D. López, Verfahren und Vorrichtung zur Innenbeschichtung von Hohlräumen durch thermisches Spritzen, German Patent: DE 102 30 847 (2002)

[4] A. Killinger, M. Kuhn and R. Gadow, High-Velocity Suspension Flame Spraying (HVSFS), a new approach for spraying nanoparticles with hypersonic speed, Surface & Coatings Technology (2006)

[5] G. Flores, H.-W. Hoffmeister C. Schnell, Vorbehandlung und Honen thermischer Spritzschichten, Jahrbuch Schleifen, Honen, Läppen und Polieren (2007)

[6] D. López, Überschallflammspritzen (HVOF) von metallurgischen und cermetischen Schichten für Zylinderlaufflächen, Ph.D. thesis, Shaker Verlag (2008)

[7] R. Gadow, A. Killinger J. Rauch, New results in High Velocity Suspension Flame Spraying (HVSFS), Surface & Coatings Technology (2008)

MULTILAYER COATINGS FOR ANTI-CORROSION APPLICATIONS

L. Lin,[1] C. Qu,[1] R. Kasica,[2] Q. Fang,[3] R. E. Miller,[1] E. Pierce,[1] E. McCarty,[4] J. H. Fan,[1] D. D. Edwards,[1] G. Wynick,[1] and X. W. Wang[1*]

1.   School of Engineering, Alfred University, Alfred, NY 14802, USA
2.   Center for Nanoscale Science and Technology, NIST, Gaithersburg, MD 20899, USA
3.   Oxford Instruments Plasma Technology, Yatton, Bristol BS49 4AP, UK
4.   Materials Technologies Consulting, LLC, Clarkston, MI 48346, USA

ABSTRACT
Due to low mass densities, magnesium alloy materials are utilized to reduce vehicular weights. However, when a magnesium alloy part is fastened with a carbon steel bolt, galvanic corrosion will occur. To stop or slow down galvanic current flow, an isolation material may be inserted between the magnesium alloy part and the steel bolt. In a previous study, an aluminum oxide thin film (100 – 300 nm) or a silicon nitride thin film (20 – 850 nm) was coated on a steel substrate as a barrier layer. In this study, besides the single layer coatings, double layer and triple layer coatings are fabricated with various combinations of the silicon nitride thin film and oxide films. After a multilayer coating is applied onto a 1050 carbon steel disc, the galvanic current between the coated steel disc and AM60B magnesium alloy is obtained. Using the lowest current value as the criterion, a candidate multilayer stacking design is then selected for the coating on the carbon steel bolts. The coated bolts are subsequently evaluated by a salt spray testing method. So far, the best layer stacking has three layers in the following sequence: silicon nitride base layer, aluminum oxide and UV curable oxide topcoat.

I. INTRODUCTION

To reduce vehicle weight, the automotive industry is currently using lighter materials, such as magnesium alloys. However, when such alloy materials are in contact with steel fasteners, galvanic corrosion takes place. Previous studies provided some guidelines for the understanding of the galvanic corrosion problems.[1-2] In Reference 1, galvanic corrosion of magnesium materials was systematically studied. The results of the study indicated that the galvanic current density distributions caused by the interaction of two independent galvanic couples can be treated as linear superposition of individual galvanic current density distribution caused by each galvanic couple. That is, the most meaningful sample for the galvanic corrosion study is a single galvanic couple such as a magnesium-steel pair. However, the majority of the results in Reference 1 were related to the metallic samples without any film coatings. Thus, it is important to study the galvanic corrosion barrier behavior after a metal piece is coated with a film. In Reference 2, an oxide film was coated on a magnesium piece via the Plasma Electrolytic Oxidation (PEO) process. The oxide film did provide certain resistance to the general corrosion.

*Corresponding author: fwangx@alfred.edu

However, the majority of the results in Reference 2 were not related to the "direct" galvanic corrosion measurements. Thus, it is desirable to study the galvanic corrosion barrier behavior when a film is coated onto one metal piece in a galvanic couple, known as the coating on the cathode. There are several film materials to be considered, such as metals, polymers and ceramics. Our focus has been on the ceramic film materials. As an example, the silicon nitride thin film coating exhibited a much higher impedance modulus value (up to 3 orders of magnitude) than that of the control substrate.[3] A ceramic film with high impedance value should act as a barrier layer to stop or slow down the galvanic current flow. As another example, a combination of an aluminum oxide thin film and a UV curable oxide coating had the impedance modulus value higher than that of the individual coating.[4] In this study, the anti-galvanic corrosion behavior of a multilayer coating is investigated.

II. EXPERIMENTS

Three coating materials are selected: silicon nitride, aluminum oxide and UV-curable oxide materials. The silicon nitride (SiN) thin films are fabricated by a plasma deposition technique, with the film thickness varying from 20 nm to 850 nm.[5] The aluminum oxide (AlO) thin films are fabricated by an Electron beam evaporation technique with or without an oxygen ion beam, with the film thickness varying from 100 nm to 300 nm.[6] The thickness selection is based on the previous results and/or the application requirements.[5-6] If the thickness is thinner than 100 nm, the corrosion barrier cannot be formed. If the thickness is thicker than several micro-meters, the coatings on the fasteners may cause the problems for the mechanical tolerances when the bolts are in contact with the nuts. Three types of the UV curable oxide films are fabricated (UV-oxide). For the films containing CeO_2 materials (UVCeO), the precursors are $Ce(OCH_2CH_2OCH_3)_3$ and a reaction initiator called AR Base.[7] For the films containing Al_2O_3 (UVAlO), the precursors are $(Al(OCH_2CH_2OCH_3)_3)$ and AR Base.[7] In addition, a mixture of UVCeO and UVAlO is also fabricated, with the AlO:CeO ratio varying from 3:1 to 1:3 (UVmix). Different combinations of films are coated onto the steel substrates as illustrated in Table I. For a double layer or a triple layer coating, the left hand side is the "bottom coating" on the carbon steel substrate, and the right hand side is the "top coating."

Table I.   Combinations of Coatings

Single Layer	Double Layer	Triple Layer
SiN	SiN|AlO	SiN|AlO|UVAlO
AlO	SiN|UVAlO	SiN|AlO|UVCeO
UVAlO	SiN|UVCeO	SiN|AlO|UVmix
UVCeO	SiN|UVmix	
UVmix	AlO|UVAlO	
	AlO|UVCeO	
	AlO|UVmix	

The galvanic current measurement, known as the zero resistance ammeter (ZRA) measurement, is carried out with Solartron 1287 Potentiostat.[8] The data are processed with a software package called CorrWare.[8] At two ends of a rectangular salt bath container, a magnesium alloy plate (AM60B) and a 1050 carbon steel disc[9] are placed respectively, with the distance between two metals being 67 mm. For the magnesium plate, the length and width are approximately the same. (The nominal dimension is 38 mm X 38 mm X 4 mm.) Before the ZRA measurement, the surface of the magnesium plate is polished with a sandpaper (grit 200), washed with IPA and dried with the compressed air. With Gamry masking tapes, the exposure area of the magnesium plate is defined as 100 mm^2 on the side facing the steel disc. The diameter of the carbon steel disc is 38 mm, and the thickness is 3 mm. The uncoated carbon steel disc surface is prepared with the same procedure as that of the magnesium plate. The exposure area of the steel disc is 100 mm^2 as defined by Gamry tapes. After salt water[10] is poured into the container, ZRA measurement is recorded. Usually, after 20 minutes or so, the current reading is relatively stable. For this study, the galvanic corrosion current is the value obtained at 30 minutes. The current density for the control is approximately 883 mA/mm^2. Among all single layer coatings, the silicon nitride coated steel disc has the lowest current density, 79 mA/mm^2. With a triple layer (SiN|AlO|UVAlO) coating, the current density is reduced to 55 mA/mm^2. As for the double layer coatings, the current densities are usually between the value of the control and that of the triple layer as illustrated in Table II.[11]

Table II. Corrosion Currents

Sample ID	Corrosion Current (mA/mm^2)
Control	883
SiN	79
AlO	812
UVAlO	805
SiN|AlO	120
SiN| UVAlO	110
AlO|UVAlO	695
SiN|AlO|UVAlO	55

A galvanic corrosion testing assembly is illustrated in Figure 1, which consists of a nylon block (purple outline), a magnesium plate (light grey), and a Kamax M8 flange head carbon steel bolt.

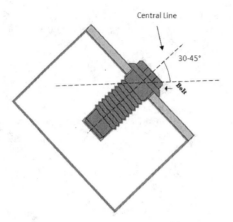

Figure 1.   Tilted sample assembly

To clean the surface of the magnesium plate (AM60B), an abrasive wheel is utilized. To examine the peak-to-valley distance due to the polishing media, Zygo 3D optical surface profiler (NewView 700) is utilized. The distance is estimated to be a few micrometers. To fasten a bolt onto the nylon block, a torque less than 3 N-m is applied. The assembled setup is illustrated in Figure 2 (A), with an uncoated M8 steel bolt on the magnesium plate before the salt mist spray testing.[12] The sample assembly is then placed in a salt mist spraying chamber, with a titled angle between 30 and 45 degrees shown in Figure 1. (For each sample to be discussed, the arrangement is the same as that illustrated in Figure 1.) After closing the chamber, salt mist is generated by an ultrasonic nebulizer from a salt water container,[13] and fed into the chamber for 15 minutes every hour. To take photos and to rotate the samples, the chamber is opened briefly every day. A typical testing duration is 12 days. With the salt mist spray testing, the uncoated carbon steel bolt and the magnesium plate are first evaluated to establish the base line. In Figure 2 (B), a photo is illustrated for the assembly with the bolt head and the magnesium plate after 12 day testing. The corrosion on the top surfaces of the steel bolt head and the magnesium plate is clearly visible. The brown corrosion spots on the steel bolt head are due to general corrosion, and the white/grey/dark corrosion spots near the interface between the steel bolt head and the magnesium plate are due to galvanic corrosion. The dark spots on the magnesium plate are due to general corrosion. The bolt is then removed to expose the contact area under the bolt head and near the interface between the bolt head and the magnesium plate, as shown in Figure 2 (C). In Figure 2 (D), the contact area on the magnesium plate is further exposed after removing the corrosion products via sandblasting. A galvanic corrosion "ring" is now visible. To highlight the "ring," two circles are added in Figure 2 (E). The average depth of the ring is approximately 2.0 mm, and the average width is approximately 1.7 mm.

(A) Bolt head on Mg plate before salt spray testing

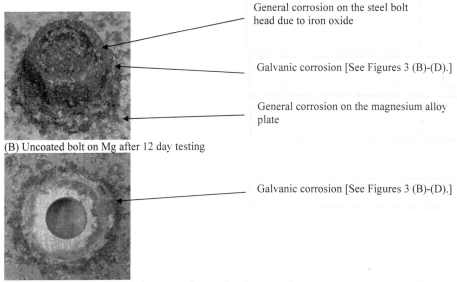

General corrosion on the steel bolt head due to iron oxide

Galvanic corrosion [See Figures 3 (B)-(D).]

General corrosion on the magnesium alloy plate

(B) Uncoated bolt on Mg after 12 day testing

Galvanic corrosion [See Figures 3 (B)-(D).]

(C) After the bolt is removed, the magnesium surface is exposed.

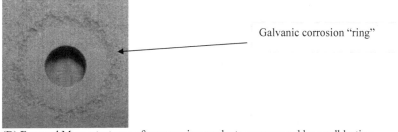

Galvanic corrosion "ring"

(D) Exposed Mg contact area after corrosion products are removed by sandblasting
Figure 2 (A)-(D) Assembly before testing, after testing with magnesium surface exposed

Figure 2 (E) Highlighted "ring"

In Figure 3 (A), an assembly with a magnesium plate and a single layer aluminum oxide coated steel bolt is illustrated before the salt mist spray testing. The surfaces of the magnesium plate and the coated bolt head appear to be shiny. In Figure 3 (B), the assembly after 12 day testing is illustrated. In comparison with the control in Figure 2 (B), the galvanic corrosion is less visible. After the bolt being removed, the exposed area of the magnesium plate is illustrated in Figure 3 (C). After sandblasting, the exposed surface of the magnesium plate is shown in Figure 3 (D). The galvanic corrosion is less visible than that illustrated in Figure 2 (D).

Figure 3 (A) Bolt coated with aluminum oxide on magnesium plate

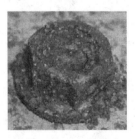

Figure 3 (B) Compared with Figure 2 (B), the corrosion is less visible here.

Figure 3 (C) After the bolt is removed, the magnesium surface is exposed.

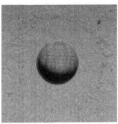

Figure 3 (D) Exposed area after sandblasting

In Figure 4 (A), another assembly with a magnesium plate and a single layer UVmix coated steel bolt is shown before the spray testing. The circular lines are due to the mechanical abrasive wheel polishing. After testing, the assembly is shown in Figure 4 (B). The galvanic corrosion is less visible than that in Figure 2 (B). After the bolt is removed, the exposed magnesium surface is shown in Figure 4 (C). After sandblasting, the exposed surface is shown in Figure 4 (D). The galvanic corrosion is less visible than that in Figure 2 (D).

Figure 4 (A) Bolt coated with single layer UVmix on Mg plate

Figure 4 (B) Compared with Figure 2 (B), the corrosion is less visible here.

Figure 4 (C) After the bolt is removed, the magnesium surface is exposed.

Figure 4 (D) Exposed area after sandblasting

In Figure 5 (A), an assembly with a magnesium plate and a triple layer SiN|AlO|UVmix coated steel bolt is shown before the testing. The circular lines are due to the mechanical abrasive wheel polishing. In Figure 5 (B), the assembly is shown after the testing. In Figure 5 (C), the exposed area of the magnesium plate is shown after the bolt is removed. Compared with Figure 2 (C), the corrosion is less visible here. In Figure 5 (D), the exposed magnesium surface is shown after the sandblasting.

Figure 5 (A) Triple layer coated bolt on Mg

Figure 5 (B) Compared with Figure 2 (B), the corrosion is less visible here.

Figure 5 (C) After the bolt is removed, the magnesium surface is exposed.

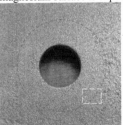

Figure 5 (D) Magnesium plates after sandblasting

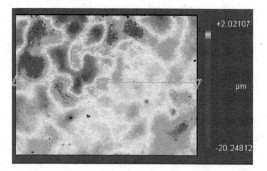

Figure 5 (E) Zygo surface profile of the area outlined by the yellow box in Figure 6 (D)

As illustrated in Figure 5 (D), there is no visible galvanic corrosion near the interface between the coated steel and the magnesium plate, as highlighted by a circle. (The general corrosion spots on the magnesium plate are due to the salt spry directly on the magnesium material.) In Figure 5 (D), a yellow rectangular box is indicated for the Zygo surface profiler measurement shown in Figure 5 (E). The topological difference is only a few micrometers, which may be due to several reasons such as the mechanical abrasive wheel polishing, sandblasting, and/or corrosion. More detailed studies will be carried out with better mechanical cleaning/polishing and chemical removal of the corrosion products (without sandblasting). As for the two layer coated bolts, the galvanic corrosion patterns are usually less visible than those illustrated in Figures 3 (D) and 4 (D); and more visible than that illustrated in Figure 5 (D). In general, the corrosion can be studied by the Zygo surface profiler and/or other surface profiler techniques. There is a one-to-one correspondence between the surface profiles (or the degree of corrosion) and the ZRA measurements.

III.   CONCLUSION AND DISCUSSION

To stop or slow down the galvanic corrosion between a magnesium alloy and carbon steel, a ceramic film is tested as the barrier coating on the steel. As far as the single layer is concerned, silicon nitride thin film should be a candidate for the further studies. The triple layer coating (SiN|AlO|UV-oxide) has a certain anti-galvanic corrosion behavior. However, the difference between the UVAlO and UVmix should be further clarified. A DOE (design of experiment) is currently being conducted, which will provide the needed information for the next experiment. The new experiment will use AZ31 magnesium alloy plates and will evaluate several parameters such as film thickness, layer stacking sequence, number of layers and film fabrication conditions. In addition, more precise measurements will be carried out to accurately calculate the mass loss due to the galvanic corrosion. As for the general corrosion of the magnesium materials, we are currently conducting a systematic study, which may be important in some applications.

ACKNOWLEDGEMENT
Guidance from USCar-USAMP member companies is greatly appreciated. In particular, suggestions and comments from Mr. Jim Quinn, Mr. Dick Osborne and Dr. Yar-ming Wang are very helpful. Additional help from Ms. Martha Sanford-Curtin, Ms. Zheng-Ying Wang and Mr. Muqiong Liu is also greatly appreciated. The partial financial support from USCAR-USAMP is acknowledged.

REFERENCES
1.   J. Jia, G. Song, and A. Atrens, "Influence of Geometry on Galvanic Corrosion of Az91d Coupled to Steel," *Corros. Sci.,* 48 [8] 2133-53 (2006).
2.   P. Zhang, X. Nie, and D.O. Northwood, "Influence of Coating Thickness on the Galvanic Corrosion Properties of Mg Oxide in an Engine Coolant," *Surf. Coat. Technol.,* 203 [20-21] 3271-7 (2009).
3a. Y. Liu, C. Qu, R. E. Miller, D. D. Edwards, J. H. Fan, P. Li, E. pierce, A. Geleil, G. Wynick, and X. W. Wang, "Comparison of Oxide and Nitride Thin Films -

electrochemical impedance measurements and materials properties," Ceramic Transactions, Vol. 214, 2010, pp. 131-146.

3b.   Y. Liu, "Silicon Nitride and Other Oxide Thin Film on Carbon Steel Substrates - Anti-Corrosion Barrier Layer Application"; M.S. Thesis. Alfred University, Alfred, NY, 2008.

4a.   C. Qu, P. Li, J. H. Fan, D. D. Edwards, W. Schulze, G. Wynick, R. E. Miller, L. Lin, Q. Fang, K. Bryson, H. Rubin, and X. W. Wang, "Aluminum Oxide and Silicon Nitride Thin Films as Anti—Corrosion Layers, Ceramic Engineering and Science Proceedings, Vol. 31, No. 3, pp. 123-134, 2010.

4b.   C. Qu, "Aluminum Oxide Thin Films as Anti-Corrosion Layers"; M.S. Thesis. Alfred University, Alfred, NY, 2009.

5.   The plasma systems include Oxford plasma ALD system, and PECVD systems such as Trikon/Aviza Plannar Fxp System, Plasma-Therm 790+ System and the Advanced Vacuum Vision 310 System.

6.   Vacuum Process Technology (VPT), Plymouth, MA 02360

7.   The UV-Curable materials were purchased from Chemat Technology, Northridge, CA 91324.

8.   Solartron Analytical (AMETEK Advanced Measurement Technology, Hampshire, UK.

9.   Muthig Industries, Fond du Lac, WI 54935

10.   To prepare the solution, 5 g of NaCl is mixed with 95 g of distilled water.

11.   For SiN|UVAlO coating, the measured corrosion current is relatively high, which may be related to the UVAlO coating fabrication.

12.   The resistance between the steel bolt head and the magnesium plate is approximately 200,000 Ohms.

13.   To prepare the salt solution, 5 g of NaCl is mixed with 100 g of distilled water.

Environmental Barrier Coatings for Turbine Engines and Extreme Environments

PLASMA SPRAY-PHYSICAL VAPOR DEPOSITION (PS-PVD) OF CERAMICS FOR PROTECTIVE COATINGS

B. J. Harder and D. Zhu, NASA Glenn Research Center, Cleveland OH 44135

ABSTRACT

In order to generate advanced multilayer thermal and environmental protection systems, a new deposition process is needed to bridge the gap between conventional plasma spray, which produces relatively thick coatings on the order of 125-250 microns, and conventional vapor phase processes such as electron beam physical vapor deposition (EB-PVD) which are limited by relatively slow deposition rates, high investment costs, and coating material vapor pressure requirements. The use of Plasma Spray – Physical Vapor Deposition (PS-PVD) processing fills this gap and allows thin (< 10 m) single layers to be deposited and multilayer coatings of less than 100 m to be generated with the flexibility to tailor microstructures by changing processing conditions. A PS-PVD processing facility has been recently built at the NASA Glenn Research Center and is being applied to the development of advanced coatings for turbine engine hot section components. To develop a basis for understanding the range of processing parameters and the effect on coating microstructure for this new processing capability at NASA GRC, a design-of-experiments was used to deposit coatings of yttria-stabilized zirconia (YSZ) onto NiCrAlY bond coated superalloy substrates to examine the effects of process variables (Ar/He plasma gas ratio, the total plasma gas flow, and the torch current) on chamber pressure and torch power. Coating thickness, phase and microstructure were evaluated for each set of deposition conditions. Low chamber pressures and high power were shown to increase coating thickness and create columnar-like structures. Likewise, high chamber pressures and low power had lower deposition rates, but resulted in flatter, more homogeneous thicknesses. The trends identified in this study are being used to improve coating processing control and to guide parameters for tailoring the microstructure of advanced coatings.

INTRODUCTION

Thermal and environmental barrier coatings (TBCs, EBCs) are necessary for the protection of metal and ceramic components in high temperature turbine engine environments. To meet ever-increasing temperature demands for improving efficiency, these coatings have become increasingly complex in composition and architecture. Thermal spray technology has been a longstanding processing method for depositing TBCs and EBCs.[1] In traditional air plasma spray (APS), coatings are formed by the buildup of molten ceramic material as a torch traverses the substrate. Electron beam-physical vapor deposition (EB-PVD) has been used to create smooth and more strain tolerant TBCs by growth of columnar grains. In this process, vaporized ceramic material condenses onto the hot substrate surface under high vacuum conditions (< 0.01 Pa). The resultant coating has a microstructure that is well suited for turbine airfoil components.[2] However, these processes both exhibit limitations, as APS methods produce rough coatings on the order of 125-250 microns thick (a single pass is 50-80 microns), and conventional EB-PVD processing is limited by slower growth rates, high equipment investment costs, and coating material vapor pressure requirements.

A new processing technology, known as Plasma Spray – Physical Vapor Deposition (PS-PVD) has been developed in order to bridge the gap between conventional APS and PVD techniques to create unique microstructures.[3,4] The PS-PVD system at the NASA Glenn Research Center is one of only five systems currently available at this time. Conventional low pressure plasma spray (LPPS) is a variation of traditional plasma spray by which deposition is done in a controlled environment at a reduced pressure, typically 6-27 kPa. PS-PVD (also known as Very Low Pressure Plasma Spray (VLPPS) or Low Pressure Plasma Spray–Thin Film (LPPS-TF)) is a unique type of LPPS where pressures are considerably lower, ranging from 60-1400 Pa. The low operating pressure results in a plasma that can extend over 2 meters in length and 1 meter in diameter. The process produces supersonic gas streams (~2000 m/s) with temperatures in excess of 6000K. Although the plasma is significantly larger than in standard vacuum plasma spray techniques, there is a more uniform distribution of temperature and particle velocity.[5] The high power levels (>100kW) at pressures below 1333 Pa result in the vaporization of the injected ceramic powder and deposition onto the target substrate. This technology has been reported to coat an area greater than 0.5 m^2 with a 10 m layer of Al_2O_3 in less than one minute.[4] The incorporation of the evaporated material into the gas stream also provides some non line-of-sight coverage, which is not possible using PVD or traditional plasma spray methods.[6]

The advantage of the PS-PVD technique is the flexibility to vary coating architecture with the processing conditions. Previous work has shown that thin, dense coatings can be deposited with splat-like microstructure as well as columnar, PVD-like coatings using the same chamber.[6-8] The wide range of possible microstructures and fast deposition rates make this technology attractive for a wide range of applications, including wear or electrically resistive coatings, diffusion barrier layers, ion-transport layers for fuel cell components, or gas sensing membranes.[4] Due to the distinctive differences between PS-PVD and traditional plasma spray deposition, operating conditions and process-properties relationships are expected to be unique and therefore must be explored. This work will focus on how variation of the processing conditions can be used to create different microstructures with a single material using PS-PVD. The resultant coatings will not be optimized for any type of durability or structure. Although the authors will not comment on properties such as thermal conductivity, density, durability, and erosion resistance at this point in time, they will be measured and discussed with respect to this study in the future. Here we will discuss the relationship between arc current, plasma gas ratio and total plasma gas flow on processing conditions, as well as the resulting coating phase and microstructure.

To showcase this emerging technology, we have chosen to work with yttria-stabilized zirconia (YSZ) as a model system. The literature has many examples of both splat (via APS processing) as well as columnar (via EB-PVD processing) microstructures for YSZ.[1,9] Although this material is traditionally used for thermal barrier coatings, this study will assist in understanding process-microstructure relationships, which will be used in developing more advanced turbine EBCs for ceramic matrix composites where high temperature capability is required.

EXPERIMENTAL PROCEDURE

Samples (25.4 mm diameter, 3.18 mm thick) of Rene N5 and Inconel 738 superalloy substrates were grit blasted and a NiCrAlY (Ni-22Cr-10Al-Y) bond coat was applied via LPPS at APS Materials, Inc. (Dayton, OH). The backsides of the samples were spot-welded to stainless steel strips, which were in turn spot-welded to a 127 x 127 mm^2 Inconel 716 plate. Two samples were mounted on the plate at a time, side by side and approximately 6.35 mm apart. The plate was mounted on the sample arm within the chamber, and set at a distance of 1650 mm from the plasma torch. The plasma torch used was a 03CP (Sulzer Metco Inc., Westbury, NY) torch with a MultiCoat processing center, and the sample arms were controlled by a GE Fanuc CNC (Hoffman Estates, IL).

Coatings were deposited from the same batch of material; a yttria stabilized zirconia ($Zr_{0.92}Y_{0.08}O_{1.96}$) Sulzer Metco (Westbury, NY) 6700 powder specifically developed for vapor deposition. The powder was spray dried and sintered to a particle distribution d10, d50, and d90 of 2.5, 6.1, and 22.0 microns, respectively. The powder is delivered to the torch via a Sulzer Metco Inc. (Westbury, NY) Model SMW 60C powder feeder. The unit uses a rotating disc speed (rpm) to regulate the powder feed rate. Current was delivered to the torch from two 1000A TriStar (EI

Sugundo, CA) power supplies. To control the pressure in the chamber, two vacuum pumps are downstream with a blower, which controlled the pressure by setting the motor speed (rpm).

Prior to starting the torch, the chamber was evacuated to <133 Pa, and then backfilled with Ar to ~4 kPa. The plasma torch was lit, and the chamber was pumped down to a pressure of ~400 Pa. The torch and sample arm were moved into position, and parameters were adjusted to preheat the substrate. The conditions for the preheating treatment are shown in Table I. After this step, the processing setpoints for the YSZ deposition condition were set and achieved within 15 seconds, at which point the powder feeding was initiated. The pressure during the 5 minute deposition was measured by a transducer inside of the deposition chamber, and the average values during the powder feeding were reported.

Table I: Preheating conditions for all experiments.

Parameter	Value
Current (A)	1400
Ar/He Ratio	0.8
Total Plasma Gas Flow (NLPM)	90
Sample-to-torch distance (mm)	1650
Heating Time (s)	60

For the design of experiments in this study, three processing variables were examined. The torch current, the Ar/He plasma gas ratio, and the total plasma gas flow were analyzed at three levels, which are shown in Table II. Many other processing conditions remained constant for all runs. Carrier gases to both powder injection ports were set at 10 NLPM (Normal Liters Per Minute) each, the sample-to-torch distance was held at 1650 mm, the powder disc speed was set to 1 rpm (~0.3 g/s), the blower speed was held constant at 1200 rpm, and deposition time was 5 minutes. Design-Expert Version 8.0.1 software (Stat-Ease, Minneapolis, MN) was used to generate the run order and analyze how the input parameters affect the chamber pressure and deposition power.

Table II: Varied conditions for the design of experiments.

Parameter	Value
Current (A)	1400, 1600, 1800
Ar/He Ratio	1:2, 1:1, 2:1
Total Plasma Gas Flow (NLPM)	80, 100, 120

After deposition, the coatings were X-rayed for phase ID using a Bruker D8 Advance diffractometer (Madison, WI). After X-raying the surface, samples were cut, mounted, and polished to a 1 micron diamond finish. Thicknesses were measured for all of the coatings using a Zeiss (Oberkochen, Germany) optical microscope. Average thicknesses were determined over a 500 micron distance, and seven such areas were chosen randomly across the entire cross-section. The average value was used as the coating thickness for the particular set of conditions.

PROCESS MODELING

The PS-PVD process is a new technique, and the facility at NASA-Glenn is a custom-built system. Therefore, it is essential that the processing conditions are well established and understood. While many process variables can be controlled, understanding the process through the plasma and at the substrate is the most effective way to control the coating deposition. Since this is a custom system, the effect of process variables on deposition conditions is not fully understood. Therefore, a statistical model was utilized to understand the influence of Ar/He plasma gas ratio, total plasma gas flow, and torch current. The effect of these processing variables was measured by chamber pressure and torch power, which will be discussed as the deposition conditions. Both chamber pressure and torch power are expected to be highly influential on substrate temperature and growth rate. Low chamber pressures increase the plasma length and width, which better envelops the target. Elevated power conditions will increase the overall plasma temperature, which may heat the substrate more effectively. Unfortunately due to grounding effects from the chamber/sample arm and the lack of a pyrometer in the chamber, it was not possible to accurately measure the temperature during coating deposition, but visual observation and measurement of the substrate temperature (with a pyrometer) after deposition supported these assumptions.

To understand and quantify the influence of the processing variables (Ar/He ratio, total plasma gas, current) on chamber pressure and torch power, a design of experiments was utilized. Our experimental design called for 3 variables to be examined at 3 levels. To evaluate the deposition conditions as a function of the process variables, a quadratic model was considered to encompass the first and second order effects, as well as any two-way interactions. In a traditional (full-factorial) experimental matrix of 3 variables at 3 levels, 27 experiments are required to separate all first and second order effects, as well as any two-way interactions between the variables, not including replicants. To minimize the number of experiments needed, we employed a d-optimal strategy to limit the number of runs to 20, which included 5 replicants. Additional repeats (replicants) were important since they helped to determine the consistency at a single deposition condition. Terms that were not considered statistically significant (confidence >90%) were removed from the model. The run conditions and resulting deposition conditions (pressure and power) are shown in Table 3.

Table III: Summary of the data for the coatings created using the design of experiments.

Run Order	Ar/He Plasma Gas Ratio	Total Plasma Gas (NLPM)	Current (Amperes)	Chamber Pressure (Pa)	Torch Power (kW)	Thickness (microns)
1	2:1	100	1600	171	74.6	18.2
2	1:2	80	1800	144	84.6	30.2
3	1:2	80	1400	144	61.1	14.9
4	1:2	120	1400	199	67.6	11.6
5	1:1	100	1600	169	80.2	22.7
6	1:1	100	1600	169	78.9	20.6
7	1:2	100	1600	171	79.5	25.4
8	1:1	120	1600	197	82.1	13.6
9	1:1	100	1600	171	78.2	20.4
10	1:1	100	1600	172	76.9	22.7
11	1:1	100	1400	169	62.4	9.6
12	2:1	80	1400	145	57.1	12.7
13	1:1	100	1800	175	86.4	27.5
14	1:1	100	1600	172	75	15.5
15	2:1	120	1800	199	89.4	9.5
16	2:1	80	1800	148	82	25.5
17	2:1	120	1400	199	64	2.6
18	1:2	120	1800	200	92.3	18.6
19	1:1	100	1600	173	75	22.7
20	1:1	80	1600	148	71.7	26.4

Control of chamber pressure is essential to the coating process, since it influences the plasma diameter and length. Figure 1 is a plot of the chamber pressure against total plasma gas. All 20 depositions are included in the chart and the variability in the vacuum level during the deposition varied by ±4-7 Pa. For equivalent pumping conditions, the pressure was insensitive to both current and Ar/He ratio. Therefore for the experimental parameters in this study, the viscosity of the gas (which would be affected by the ratio and current) does not impact the deposition condition and the total plasma gas was found to be the only factor influencing the chamber pressure. The standard error for the empirical model was 1.73 Pa, and the R^2 value was 0.99.

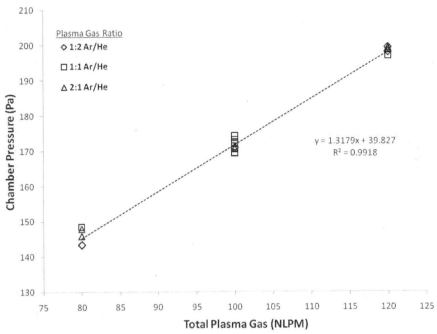

Figure 1. Plot of chamber pressure as a function of total plasma gas. The pressure was insensitive to current and gas ratio.

The torch power results in Table III were measured as the product of arc voltage and the supplied amperage. The empirical model (standard error = 1.40 kW, R^2 = 0.98) for plasma power at the two different gas mixtures is plotted in Figure 2. First order effects were observed for all three variables, as well as a second order effect from the current. A lower Ar/He ratio also served to increase the torch power, which was expected. Increasing the helium content serves to increase the plasma enthalpy and generate more heat. Higher plasma gas flow rates (total plasma gas) elevated the power by pushing the arc further out of the torch, which increased the voltage. Given equivalent current settings, higher plasma gas flow rates therefore produce a higher power plasma.

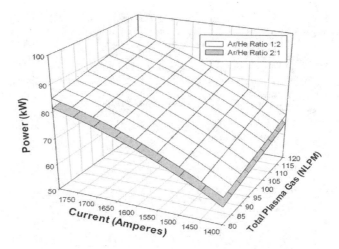

Figure 2. The empirical model surfaces of the measured torch power at Ar/He ratios of 1:2 and 2:1.

The aforementioned models provide a basic understanding of how some of the PS-PVD processing variables influence the deposition conditions. The empirical models of the deposition conditions are independent of coating material. Therefore, these relationships should hold true regardless of the coating system and substrate material. With these models, the effect of chamber pressure and torch power can be examined on the microstructure for a single coating material.

RESULTS – COATINGS

The deposition conditions discussed above have a significant effect on coating microstructure. Coating thickness, phase composition, and microstructure were examined as a function of chamber pressure and torch power. The thicknesses reported in Table III were averages of optical microscopy measurements across each sample. In Figure 3 coating thickness is shown as a function of torch power and grouped in similar pressures to examine the effect of both conditions. The chamber pressures of the run conditions in Table III fell within 3 ranges: 144-148, 172-175, and 197-200 Pa. The spread within each of these ranges was within the 4-7 Pa variability in chamber pressure during deposition. Therefore, the pressures can be grouped and analyzed by range. High power and low pressure resulted in the thickest coatings, and these trends were consistent for the 144-148 and 172-175 Pa conditions. The low growth rate and significant scatter observed at high pressures (197-200 Pa) may be a consequence of a reduced amount of vapor phase available due to the standoff distance. A reduction in the torch-to-sample distance may limit the spread in the data or reduce the effect on coating thickness, but this is outside the scope of this current study.

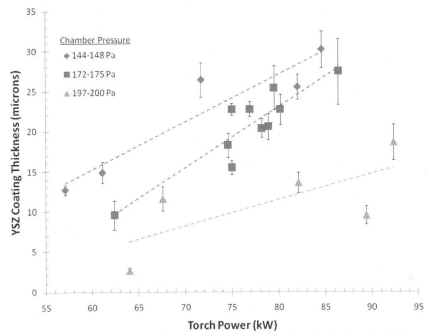

Figure 3. Plot of coating thickness as a function of torch power. The dashed lines are a linear fit of the data to show the trends for like chamber pressure conditions, and the error bars shown are the standard deviation of the coating thickness across the sample surface.

    The coatings were also X-rayed for phase identification. As-sprayed diffraction patterns for 3 growth conditions and the as-received powder are shown in Figure 4. The majority of the deposited coatings was tetragonal YSZ as expected. However, in the lowest enthalpy conditions (which were also the thinnest coatings), traces of monoclinic zirconia and yttria were observed. The starting YSZ powder was spray dried and sintered and the as-received powder (bottom curve of Figure 4) consisted of a mixture of Y_2O_3 and ZrO_2. In the PS-PVD process, the particles are vaporized and mixed within the plasma. The presence of monoclinic ZrO_2 and cubic Y_2O_3 in the top curve of Figure 4 may have resulted from inadequate mixing of the two materials during processing. Under the high enthalpy conditions, the plasma was more homogeneous in composition (effectively mixed within the plasma) and the YSZ was deposited in the tetragonal phase. This trend is consistent with the third curve from the top in Figure 4, which corresponded to the high enthalpy condition. The as-sprayed coating is nearly 100% tetragonal YSZ. Thin coatings also showed a larger signal of the NiCrAlY bondcoat (due to coating thickness) as shown in the top curve in Figure 4. The as-sprayed coatings did not exhibit obvious evidence of crystallographic texture. This is likely due to the bondcoat roughness and low substrate temperatures in the PS-PVD coatings when compared to traditional EB-PVD methods.

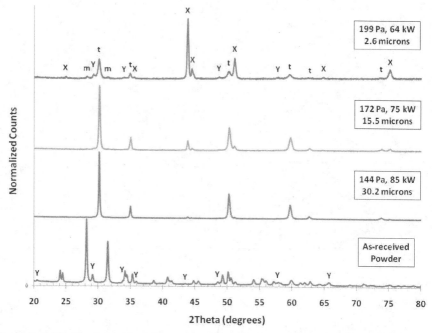

Figure 4. X-ray diffraction patterns of three different PS-PVD deposited YSZ coatings (top to bottom- experimental number 17, 10, 2) and the as-received YSZ powder. "X" designates the NiCrAlY bond coat, "Y" designates the cubic Y_2O_3 phase, "t" designates the tetragonal YSZ phase, and "m" designates the monoclinic ZrO_2 phase. Note that the thinnest coating shows the largest amount of monoclinic ZrO_2 and cubic Y_2O_3. The as-received powder is a mixture of monoclinic ZrO_2 and cubic Y_2O_3. The yttria peaks are labeled with "Y," and all other peaks correspond to the monoclinic ZrO_2.

Sample microstructures varied significantly with the deposition conditions. However, all 20 experimental conditions resulted in deposition along the sides of the substrate in addition to the surface, indicating that the YSZ was applied to some degree via the vapor phase. On average, high power, low vacuum conditions produced columnar-like coatings. High chamber pressures and low power produced flatter, more uniform layers. It was possible to observe both types of microstructures in a single sample, and they were distributed evenly across the sample cross-section. Examples of columnar-like microstructures are shown in Figure 5. The direction of column growth was perpendicular to the torch direction and substrate surface. Although some degree of deposition occurred along the sides of the substrates with the same microstructure, the surface normal to the plasma was thicker. Column widths for the highest power/lowest pressure conditions were 10-20 microns, which are wider than standard PVD growth methods, which produce columns on the order of 5-10 microns.[9] Lower power settings (and likely lower substrate temperatures) resulted in wider columns on the order of 10-20 microns. Coating thicknesses did not have any impact on column width. In Figure 5(a) and (b), the columns are tightly packed and a consistent growth is seen across the substrate surface. These images were of the coating deposited during sample run number 2, which corresponded to the low vacuum (144 Pa), high power (84.2 kW) condition. Figure 5(c) and (d) show coating microstructure deposited during run number 10, which had a higher chamber pressure (172 Pa) and a lower power (76.9 kW). The columnar-like microstructure is still present, especially on the peaks of the bondcoat. The columns began to form at the peaks in the bondcoat since it was the first point of incidence of the plasma meeting the surface. Coatings continue to grow into the columnar microstructure, and the growth rate may be enhanced by locality as well as potentially higher local temperatures. However, the growth is much less uniform than those obtained under high power condition, as the columns are spaced further apart than those in (a) and (b). In Figure 5(a)-(d), both coatings show evidence of periodic renucleation and grain growth, which indicates the substrate temperature was lower than ideal for columnar growth.[10] It would be expected that higher substrate temperatures would result in thinner, more tightly packed columns.

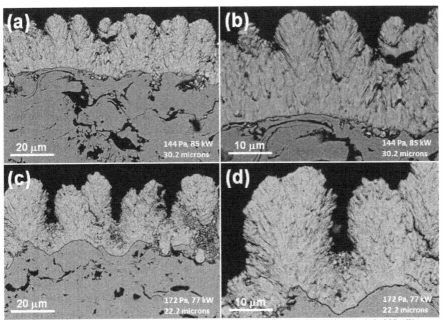

Figure 5. Backscatter images of coating cross-sections of which the light region is YSZ and the dark region is the MCrAlY bondcoat. (a) Coating deposited using conditions in run number 2 (low pressure and high power), and a higher magnification is shown in (b). (c) Coating deposited using conditions in run number 10 (model centroid), and a higher magnification is shown in (d).

Examples of flat, more homogeneous coating microstructures are shown in Figure 6. A thin (~3 micron) dense coating is seen in the micrographs in Figure 6(a) and (b), which corresponds to one of the lowest power (64 kW) and highest pressure (199 Pa) conditions. Figure 6(c) and (d) were obtained from deposition number 10 (pressure = 172 Pa, power = 76.9 kW), which was representative of the center of the design space, and were from the same sample as Figure 5(c) and (d). These images illustrate that both columnar and flat microstructures were found on a single sample. Samples with a large standard deviation in thickness (shown in Figure 3) corresponded to those where both microstructures were readily observed. The larger standard deviation is a result of a bimodal distribution in the microstructure. The coating in Figure 6(c) and (d) also shows a large amount of void space, as well as short, discontinuous columns. This porous structure suggests a 'mixed mode' of deposition. Spinhirne et al. saw a similar effect using the Low Pressure Plasma Spray – Thin Film rig located at Sandia National Laboratory, and attributed this to a substrate temperature that was too low for consistent column growth.[11] The ability to form a dense coating, as seen in Figure 6(a) and (b), may be attributed to the higher chamber pressure. At a pressure of 199 Pa, the plasma is considerably shorter, and the vaporized YSZ is more likely to cool before it reaches the substrate. A consequence of the YSZ cooling prematurely is the potential for incomplete mixing and vaporization, or condensing from a vapor into a liquid phase prior to reaching the substrate. The condensed (liquid) YSZ will impact and flatten out on the surface, or fall out of the plasma before reaching the substrate. The resulting coatings could exhibit a more traditional, splat-like coating geometry, and may not be completely tetragonal in phase due to incomplete mixing of the Y_2O_3 and ZrO_2. The microstructures in Figure 6(a) and (b) display this kind of splat-like coating, and the top pattern in Figure 4 shows the coating is not completely tetragonal in phase. However, it is not possible to know at this point in time if the coatings were solely deposited via the liquid or vapor phase, or if it was a mixed-mode process. The deposition rate for these coatings at higher chamber pressure is considerably slower, but they exhibit significantly less void space than thicker coatings.

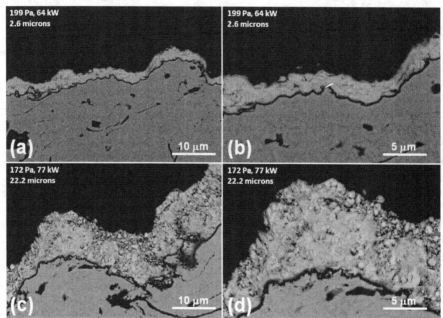

Figure 6. Backscatter images coating cross-sections of which the light region is YSZ and the dark region is the MCrAlY bondcoat. (a) Coating deposited using conditions in run number 17 (lowest power, highest pressure), and a higher magnification is shown in (b). (c) Coating deposited using conditions in run number 10 (model centroid), and a higher magnification is shown in (d).

CONCLUSIONS

The capabilities of the new processing technology, Plasma Spray-Physical Vapor Deposition (PS-PVD), were explored using yttria-stabilized zirconia as a model material. This study was undertaken to examine how some of the processing variables of the PS-PVD technology influence deposition conditions and coating microstructure. It was shown that chamber pressure scaled with total plasma gas, and that structure and coating thickness were influenced by torch power and chamber pressure, with high power and low pressure producing the columnar microstructures. As with most plasma spray technologies, the PS-PVD process is influenced by many variables, and it is not possible to fully describe the conditions for generating fully tailored microstructures with a single round of experiments. The trends observed here provide a foundation for process control and will be utilized in establishing the parameters necessary for depositing tailored microstructures. Future work will investigate additional factors such as the influence of substrate temperature and the effect of processing variables on performance.

ACKNOWLEDGEMENTS

This work was funded by the Supersonics Project under the Fundamental Aeronautics Program. The authors would also like to thank Mary Ann Meador and Rick Rogers of NASA Glenn Research Center (GRC), Joy Buehler and Mike Cuy of ASRC, NASA GRC Division, as well as Bob Pastel of SLI, NASA GRC Division, for valuable discussions and contributions to this work.

REFERENCES

[1]R. A. Miller, Current Status of Thermal Barrier Coatings - An Overview, *Surf. & Coat. Tech,* **30**[1] 1-11 (1987).

[2]A. G. Evans, D. R. Mumm, J. W. Hutchinson, G. H. Meier, and F. S. Pettit, Mechanisms Controlling the Durability of Thermal Barrier Coatings, *Prog. Mater. Sci.,* **46**[5] 505-53 (2001).

[3]A. Refke, G. Barbezat, J.-L. Dorier, M. Gindrat, and C. Hollenstein, Characterization of LPPS Processes Under Various Spray Conditions for Potential Applications, *Proc. of the Inter. Therm. Spray Conf.,* May 5-8, (Orlando, FL), 581-88 (2003).

[4]A. Refke, M. Gindrat, K. v. Niessen, and R. Damani, LPPS Thin Film: A Hybrid Coating Technology Between Thermal Spray and PVD for Functional Thin Coatings and Large Area Applications, *Proc. of the Inter. Therm. Spray Conf.,* May 14-18, (Beijing, China), 705-10 (2007).

[5]J. L. Dorier, M. Gindrat, C. Hollenstein, M. Loch, A. Refke, A. Sailto, and G. Barbezat, Plasma Jet Properties in a New Spraying Process at Low Pressure for Large Area Thin Film Deposition, *Proc of the Inter. Therm. Spray Conf.,* May 28-30, (Singapore, Singapore), 759-64 (2001).

[6]K. von Niessen, M. Gindrat, and A. Refke, Vapor Phase Deposition Using Plasma Spray-PVD (TM), *J. of Therm. Spray Tech.,* **19**[1-2] 502-09 (2010).

[7]A. Hall, J. McCloskey, D. Urrea, T. Roemer, D. Beatty, N. Spinhirne, and D. Hirschfeld, Low Pressure Plasma Spray - Thin Film at Sandia National Laboratories, *Proc. of the Inter. Therm. Spray Conf.,* May 4-7, (Las Vegas, NV), 725-28 (2009).

[8]A. Hospach, G. Mauer, R. Vaßen, and D. Stöver, Columnar-Structured Thermal Barrier Coatings (TBCs) by Thin Film Low-Pressure Plasma Spraying (LPPS-TF), *J. of Therm. Spray Tech.,* **20**[1-2] 116-20 (2011).

[9]U. Schulz, K. Fritscher, and M. Peters, EB-PVD Y2O3- and CeO2/Y2O3-stabilized zirconia thermal barrier coatings - Crystal habit and phase composition, *Surf. & Coat. Tech.,* **82**[3] 259-69 (1996).

[10]K. Wada, N. Yamaguchi, and H. Matsubara, Crystallographic Texture Evolution in ZrO_2-Y_2O_3 Layers Produced by Electron Beam Physical Vapor Deposition, *Surf. & Coat. Tech.,* **184**[1] 55-62 (2004).

[11]N. Spinhirne, D. Hirschfeld, A. Hall, and J. McCloskey, The Development and Characterization of Novel Yttria-Stabilized Zirconia Coatings Deposited by Very Low Pressure Plasma Spray, *Proc. of the Inter. Therm. Spray Conf.*, May 4-7, (Las Vegas, NV), 750-55 (2009).

AN $Yb_2Si_2O_7$ OXIDATION RESISTANCE COATING FOR C/C COMPOSITES BY SUPERSONIC PLASMA SPRAY

Lingjun Guo*, Minna Han, Hejun Li, Kezhi Li, Qiangang Fu
State Key Laboratory of Solidification Processing, Northwestern Polytechnical University
Xi'an, Shaanxi, 710072, PR China

ABSTRACT

An efficient glass coating was prepared on the surface of SiC coated C/C composites by supersonic plasma spray to prevent them from oxidation at 1873 K. The structure of the as-obtained coating was characterized by X-ray diffraction, scanning electron microscopy and isothermal oxidation test at 1873 K in air, respectively. Results show that the achieved $Yb_2Si_2O_7$ outer coating was dense and it was bonded well with the SiC inner layer such that the oxidation protective ability and the thermal shock resistance were improved. After oxidation in air at 1873 K for 328 h and thermal cycling between 1873 K and room temperature for 17 times, the mass gain rate of the coated sample was only 0.28 %. A dense layer about 5 μm between the SiC inner coating and the $Yb_2Si_2O_7$ outer coating was in favor of the enhancement of the oxidation resistance of the $SiC/Yb_2Si_2O_7$ coating for C/C composites.

Keywords: Oxidation resistance coating; C/C composites; Microstructure; Supersonic plasma spray.

INTRODUCTION

Since carbon/carbon(C/C) composites are unique in that they can maintain specific strength even above 2273 K, much effort has been spent on attempts to develop effective methods for preventing them from oxidation at high temperatures [1]. Coating is considered to be an effective way. SiC coating is one of the best bonding layers between C/C composites and the ceramic outer layer because of its good physical and chemical adaptability of coating-to-matrix and bonding layer to outer layer [2]. Pack cementation is a good method because it can provide a gradient distribution of elements at the interface between the SiC coating and the C/C composites.

Si-based non-oxidation ceramics is widely used as a coating to provide protection against oxidation owing to its excellent anti-oxidation properties and good compatibility with C/C composites. In addition, the glassy SiO2 film can be formed on the surface of the Si-based non-oxidation ceramics which can efficiently prevent oxygen from diffusing into C/C substrate in dry air [3-4]. However, the SiO2 scale will react with water vapor from combustion reactions, and forms gaseous silicon hydroxide species [5], resulting in unacceptably high recession rate of C/C composites. Therefore, the choice of outer layer materials is essential. Yb2Si2O7 has been considered as a candidate coating material for protecting SiC ceramics from high-temperature environments because of its low SiO2 activity and favorable coefficient of thermal expansion match with SiC ceramics [6-7].

So far, no literature has been published yet about using Yb2Si2O7 in the oxidation protective coating for C/C-SiC composites. Therefore, a novel Yb2Si2O7 coating was prepared on the surface of C/C-SiC composites by supersonic plasma spray In this work. And the microstructure and the oxidation protective ability of the as-received coating were primarily investigated.

* corresponding author: Guo Lingjun (1963,6—),
Tel: 086—029—88494197; E-mail: guolingjun@nwpu.edu.cn

EXPERIMENTAL

The specimens (10×10×10 mm3) used as substrates were cut from bulk 2D-C/C composites with a density of 1.70 g/cm3. After being hand-polished with No.300SiC sand papers, they were cleaned ultrasonically with distilled water and dried at 373 K for 2 h. The SiC coating was prepared by a pack cementation process with 65–80 wt.% Si, 10–25 wt.% graphite, and 5–15 wt.% Al2O3 in an argon atmosphere at 2173 ~2473K for 2 h. The preparation details were reported in [8].

The Yb2Si2O7 bulk was prepared by hot pressing method using high purity Yb2O3 and SiO2 (99.9 % purity) powders as the starting materials. Yb2O3 and SiO2 powders (A stoichiometric molar ratio of Yb2O3 and SiO2 was 1:1) was mixed in ethanol and then dried in air at 373 K for 1 day. These mixed powders were then sintered at 1723~1923 K for 12 h at ambient atmosphere in an electric furnace. The sintered bulk was milled in a ball mill to attain the Yb2Si2O7 powders with the size of 300 meshes.

The Yb2Si2O7 coating was prepared by supersonic plasma spraying the Yb2Si2O7 powders on C/C composites substrates that has been already coated with 50 to 60 μm thick SiC coating. Each surface of the sample was coated with Yb2Si2O7 in the same manner. The thickness of every Yb2Si2O7 coating was deposited about 120μm.

The isothermal and thermal cycling oxidation tests of the prepared sample were performed at 1873 K in an air and electrical furnace. The specimens were weighted before and after the cycle from high temperature to room temperature by a precision balance, and then the weight change of the samples was reported as a function of time. The structure of the coating was characterized by X-ray diffraction (XRD); the morphology and element distribution were examined using scanning electron microscope (SEM) and energy dispersive spectroscopy (EDS).

RESULTS AND DISCUSSION

Fig.1 shows the XRD patterns of the sprayed Yb2Si2O7 anti-oxidation coating samples. The Yb2O3 phase and the Yb2Si2O7 containing amorphous phase are observed. Commonly, a broad peak for amorphous silica appears around $2\theta = 15$–$30°$. The positions of the broad peaks indicate the amorphous phases are $2\theta = 27$-$35°$ and $2\theta = 50$-$65°$. Hence, it is suggested that the composition of the amorphous phase shown in Fig.1 contains Yb2O3.

All the anti-oxidation coating samples were annealed in air at 1500 ℃ for 4 hours in order to crystallize the coating phases. Fig.2 shows the XRD patterns of the coating surface after the heat treatment. The main crystalline constituents are Yb2Si2O7 and Yb2SiO5, which indicated that the amorphous phase has crystallized into Yb2Si2O7 and Yb2SiO5 phases during the post heat treatment. The difference in Yb2SiO5 phase peak intensity might be resulted from a small amount of silica component that vaporized during the plasma spray process.

Fig.3(a) displays the SEM micrograph of the coating surface. It can be found that some small pores and some micro-cracks existed on the surface. The cross section SEM image of the coating is shown in Fig.3(b). The gradient SiC coating was formed as the result of the deep infiltration of liquid Si into the coating. SiC coating and Yb2Si2O7 coating integrated by mechanical force. No gap between SiC and Yb$_2$Si$_2$O$_7$ coating is observed, implying good interface bonding. However, some pores in the Yb$_2$Si$_2$O$_7$ layer were formed in the region of stress concentration due to the bigger SiC particles.

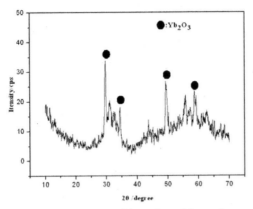

Fig.1. XRD spectrum of the surface of the coating

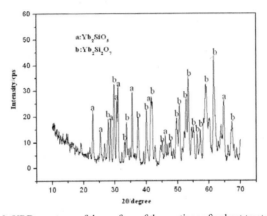

Fig.2. XRD spectrum of the surface of the coatings after heat treatment

Fig.4 is the isothermal oxidation curve of C/C composites with SiC/Yb2Si2O7 coated C/C at 1827K in air. It can be seen that the weight gain of the coated C/C composites sample is only 0.28 % after 1827 K oxidation for 328 h. The oxidation behavior of the coated sample could be divided into two processes according to the oxidation curve. When the sample was oxidized less than 50 h, its weight increased. At the initial stage of oxidation, the coating was partially oxidized and forms a limited amount of mass gain in air at 1873 K for 50 h, and the mass gain of the coated sample is mainly contributed to the oxidation of the inner SiC coating. When the sample was oxidized more than 50 h, the weight lost slowly as the oxidation time went on.

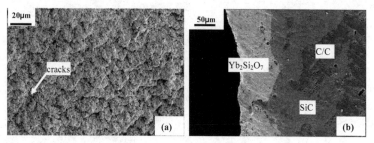

Fig.3 Surface and cross-section SEM images of SiC/Yb$_2$Si$_2$O$_7$ coating:
(a) surface and (b) cross section

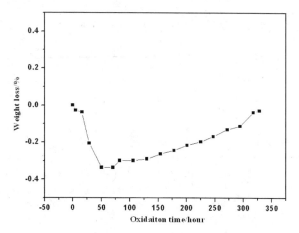

Fig.4. The isothermal oxidation curve of C/C composites
with SiC/Yb$_2$Si$_2$O$_7$ coated C/C at 1827K in air

In addition, when the sample was exposed to thermal cycling between 1873 K and room temperature for 17 times, no weight loss of the coated sample was found. So, it could be inferred that the coating has excellent oxidation and thermal shock resistance.

The surface and cross-section micrographs of the coated sample after oxidation at 1873K for 328 h are shown in Fig.5a. Some pores and cracks can be found on the Yb2Si2O7 coating surface from Fig.5(a).They might be formed during the quick cooling from 1873 K to room temperature. Moreover, a small amount of glass phase was formed on the sample surface. The pores and micro-cracks of the Yb2Si2O7 coating provided channels for the diffusion of oxygen into the inner SiC coating, and the SiO2 glass phase was generated, which results in the mass gain of the coated sample during this oxidation stage. SiO2 would diffuse into the coating surface through small pores and micro-cracks. A denser Yb2Si2O7 coating can be found in the cross-section image of the as-coated sample (Fig.5(b)).

Besides, no penetrable cracks or spallation were found after oxidation at 1873 K for 328 h and 17 thermal cycles between 1873 K and room temperature. Therefore, it could be concluded that the coating exhibits excellent thermal shock resistance. This may be attributed to the formation of the SiC gradient bonding layer and the good match of thermal expansion coefficient between the Yb2Si2O7 (3.5~4.5×10-6/℃) outer coating and the SiC (4.5×10-6/℃) internal layer [5]. A dense layer about 5 μm between the SiC inner coating and the Yb2Si2O7 outer coating was found in Fig.5(c). The molten SiO2 could not only fill the pores and micro-cracks of the coating, but also promote sintering of Yb2Si2O7. Thus, the density of Yb2Si2O7 improved. A dense layer about 5 μm between the SiC inner coating and the Yb2Si2O7 outer coating could efficiently improve the oxidation resistance of the SiC/Yb2Si2O7 coating.

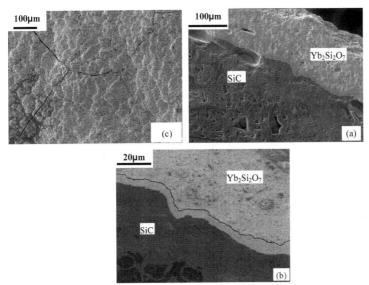

Fig.5. SEM micrographs of the coated C/C after oxidation at 1873K in air
(a) surface   (b) and (c) cross section

CONCLUSIONS

An effective Yb2Si2O7 oxidation protection coating on SiC coated C/C composites was produced by supersonic plasma spray. The as-obtained coatings have excellent oxidation protective ability and good thermal shock resistance. It could effectively protect C/C composites from oxidation for more than 328 h at 1873 K and undergo the thermal shock between 1873 K and room temperature for 17 cycles. The corresponding mass gain rate is only 0.28%.

REFERENCE

1   Morimoto T, Ogura Y, Kondo M, et al. Multilayer coating for carbon/carbon composites. Carbon 1995,33(4):351-357.

2   Zhang Yulei, Li Hejun, Fu Qiangang, Li Kezhi, et al. A Si-Mo oxidation protective coating for C/SiC coated carbon/carbon composites. Carbon 2007,45 (5):1130-1133.

3   Fu Qiangang, Li Hejun, Shi Xiaohong, et al. Silicon carbon coating to protect carbon/carbon composites against oxidation. Scripta materialia 2005,52(9):923-927.

4   Li Kezhi, Hou Dangshe, Li Hejun, et al. Si-W-Mo coating for SiC coated carbon/carbon composites against oxidation. Surface and coatings technology 2007,201(24):9598-9602.

5   Lee K N, Dennis S F, Narottam P B. Rare earth silicate environmental barrier coatings for SiC/SiC composites and Si$_3$N$_4$ ceramics. Journal of the European Ceramic Society 2005,25(10):1705-1715.

6   Su Yifeng, Lee W Y. Synthesis of silicate coating by layer-by-layer self-assembly of Yb$_2$O$_3$ and SiO$_2$ particles. Surface and coatings technology 2008,202(15):3661-3668.

7   Ueno S, Ohji T, Lin H T. Recession behavior of Yb$_2$Si$_2$O$_7$ phase under high speed steam jet at high temperatures. Corrosion Science 2008,50(1):178-182

8   Hung J F, Zeng X R, Li H J, et al. Mullite-Al$_2$O$_3$-SiC oxidation protective coating for carbon/carbon composites. Carbon 2003,41(14)2825-2829.

9   Ueno S, Ohji T, Lin H T. Recession behavior of a silicon nitride with multi-layered environmental barrier coating system. Ceramics international 2007,33(5):859-862.

Functionally Graded Coatings and Interfaces

DEVELOPMENT OF OXIDE CERAMIC APS COATINGS FOR MICROWAVE ABSORPTION

M. Floristán[1, 2], P. Müller[1], A. Gebhardt[1], A. Killinger[1], R. Gadow[1], A. Cardella[3], C. Li[4]

[1] Institute for Manufacturing Technologies of Ceramic Components and Composites (IMTCCC), Universität Stuttgart, Allmandring 7 b, D-70569 Stuttgart, Germany
[2] Graduate School of Excellence advanced Manufacturing Engineering (GSaME), Universität Stuttgart, Nobelstraße 12, D-70569 Stuttgart, Germany
[3] European Commission c/o Wendelstein 7X, Boltzmannstr. 2, D-85748 Garching, Germany
[4] Max-Planck-Institut für Plasmaphysik, EURATOM Assoc., Wendelsteinstraße 1, D-17491 Greifswald, Germany

ABSTRACT

Thermonuclear fusion is a promising source of clean energy for the future. The extreme operating conditions of fusion reactors have lead to an increasing interest on the field of high performance materials. Research has focused on the development of materials which can function under thermal and mechanical loads or strong radiation. The present work describes the development of coating systems acting as absorbers for 140 GHz radiation on the water-cooled baffle for the stellarator Wendelstein 7-X.

Several types of ceramic coatings were applied on copper substrates by Atmospheric Plasma Spraying. Different powders in composition and grain size were used as feedstock material. The influence of the process parameters on the coating properties and microwave absorbing capability was analyzed. The coatings mechanical properties were characterized in terms of porosity, microhardness, roughness, adhesion and residual stresses. XRD and SEM were carried out. It was found that thickness and microstructure of the coatings have a significant influence on microwave absorption behaviour. For Al_2O_3/TiO_2 coatings, absorption values over 90% were obtained.

After optimization of the coating structure for maximum microwave absorption, the coating procedure had to be adapted to the complex water baffle geometry. With this aim, the robot kinematics was designed to achieve a regular coating regarding thickness and microstructure.

INTRODUCTION

Fusion has the potential of providing an essentially inexhaustible source of energy for the future [1]. The principle of nuclear fusion is the reaction of two light nuclei to form a heavier nucleus releasing energy. In order to fuse, two nuclei need to have enough kinetic energy to overcome the repelling Coulomb force between them and approach each other sufficiently. This can be achieved by confining and heating the fusion fuel up to high temperatures in which the gas exists in the plasma state [1]. Among the fusion reactions, the least difficult to initiate on Earth [2] and the most prospective is the fusion of deuterium and tritium (D-T), which produces helium and a neutron: $D + T \rightarrow He + n$ [3]. The product of this fusion reaction, Helium-4, as well as hydrogen species and trace amounts of impurity elements, lead to a high contamination of the plasma in fusion reactors and have to be exhausted from the plasma vessel. Cryopumps can satisfy these needs pumping gases by cryogenic condensation [4].

Wendelstein 7-X (W7-X) is a fusion reactor of type stellarator under construction in Greifswald, Germany [5]. The cryo vacuum pumps in W7-X have a panel with supercritical He at an inlet temperature of 3.4 K [6]. The gases, which have to be exhausted from the plasma vessel, freeze at the surface of this panel and are kept bounded until the cryopump is regenerated. In order to protect the He cryopanel from the plasma radiation and a possible heating, which could hinder its operation, a water-cooled baffle is used [3]. The water-baffle is composed by several copper lamellas or chevrons, which absorb the radiation from the plasma. However, the W7-X cryopumps are expected to be subjected to high microwave radiation from the electron cyclotron radio frequency heating system, and such stray radiation is not

93

sufficiently reduced by the water cooled Cu chevrons. Therefore, the use of microwave absorbing ceramic coatings applied on the chevrons was analysed and investigated in order to improve the performance of the water baffle and thus, reduce the thermal load on the He panel of W7-X.

The use of Al_2O_3/TiO_2 thermally sprayed coatings for diverse applications on components of fusion reactors has been reported and discussed in the literature [3, 7-10]. However, scarce information is found about the optimal processing of these coatings regarding their absorption capability and about the influence of the powder and phase composition on the functionality of the coatings. Al_2O_3/TiO_2 thermally sprayed coatings were developed and optimized in this work to be deposited on water baffles.

MATERIALS AND METHODS

Thermal spraying are a group of processes in which finely divided metallic and non-metallic materials are deposited in a molten or semi-molten state on a prepared substrate [11, 12]. Atmospheric Plasma Spraying is a flexible and cost effective coating process characterised by the high reached temperatures, over 8000 K [13], which allow the processing of almost any material. By means of a high voltage electrical discharge, an arc is created between a water-cooled copper cylindrical anode and a thoriated tungsten cathode situated inside the anode, see Fig. 1 (left). A gas mixture is injected in the torch. The interaction of the electric arc with the gas mixture makes the gas atoms dissociate and ionize, leading to the formation of the plasma jet or plasma plume [11]. The spray powder is injected in the plasma. The particles are totally or partially melted by the plasma jet and propelled towards the substrate surface. Upon impacting at the surface, the particles deform and rapidly solidify building up the coating.

Planar copper substrates (50 x 50 x 2 mm^3) were coated to determine the optimal spray parameters for the application. After coating optimization on planar substrates, the spraying process was applied to a mock-up composed by Cu chevrons as prototype of the water baffle, see Fig. 1 (right). The Cu substrates were degreased and, in order to improve mechanical adhesion of the coating, their surface was roughened by grit blasting with 250 μm alumina grit, pressure of 0.6 MPa at an angle of 90° and distance of approximately 200 mm. The F6 Atmospheric Plasma Spraying torch (GTV, Germany) was used in this study. A six axis robot (Type RX 130 B, Stäubli Tec-Systems GmbH, Germany) was used to guide the spray torch and it described a meander movement to coat the samples.

Particle morphology and size distribution of all the powders used for the coating experiments were analyzed using a LEO 483 VP scanning electron microscope (Carl Zeiss SMT AG, Germany) and a laser diffraction particle size analyzer (Mastersize S, Malvern Ltd, UK), respectively.

The cross sections of the samples for microscopy analysis were cold-embedded in epoxy resin and diamond polished with RotoPol31 Struers® equipment (Struers A/S, Germany). A LEICA™ MEF4M (Leica Microsystems CMS GmbH, Germany) inverted optical microscope was used for image analysis of the coatings cross section. Coating porosity was determined by digital image analysis [14] on polished sections (Software AQUINTO, Analysis Software, Olympus Deutschland GmbH, Hamburg).

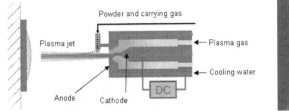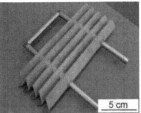

Fig. 1: Principle of atmospheric plasma spraying (left), mock-up prototype to coat (right).

Microhardness HV 0.1 was measured on coating cross section by an automated universal hardness indenter equipment Fischerscope™ HCU (Helmut Fischer GmbH, Sindelfingen-Maichingen, Germany) according to DIN 14577-1. The values given are the average of 12 measurements per sample. Adhesion measurements of the coatings were made with a universal testing machine (Zwick GmbH & Co, Germany). Pull-off tests were carried out in which a stamp or stub was affixed by an adhesive to the coating and an increasing load was applied to the stub-coating surface, until the stub was pulled off. The given results are the average of six measurements per sample.

Phase composition analysis of powders and coatings was assessed in the present study by X-ray diffraction using a Bruker AXS D8 Advance diffractometer (Bruker AXS Inc., USA) with Cu-Kα radiation. Phase identification was made with the Diffrac[Plus] Eva software superposing experimental XRD results with the ones of the software data base.

Residual stress measurements were carried out in the sprayed coatings by using the incremental hole milling and drilling method, for further details refer to [15]. Drilling diamond tools of 0.9 mm diameter and strain gauges DMS CEA-06-062UM120 (0°, 45°, 120°) from Vishay instruments were used. The drilling hole diameter was 1.8 mm and the drill depth was 500 μm with a 10 μm incremental drilling step.

In order to measure the thermal expansion coefficient of some coatings, mechanical dilatometry was used. Measurements were carried out under atmospheric conditions with a DIL 402C dilatometer (Netzsch-Gerätebau GmbH, Germany), at a heating rate of 5 K/min starting at room temperature up to 1000°C.

All the samples were characterised regarding their microwave absorption behaviour as described in [15]. The measurements were taken with parallel and perpendicular polarisation, and in each case with incidence angles of 20°, 45° and 60° and a frequency of 140 GHz.

RESULTS
Analysis of the spray angle on coating microstructure and functionality

Spray angle is the angle between the centre axis of the spray torch and the surface of the substrate, in the plane orthogonal to the torch displacement [17]. For standard applications in thermal spraying, the spray angle is usually kept close to 90°, but it will vary if the substrate has a nonplanar surface [18], as it is the case considered in this work. Due to its complex geometry, the water baffle cannot be coated with the spray torch continuously perpendicular to the chevrons surface. Instead, spray angles lower than 90° are necessary to reach with the plasma jet the complete surface of the chevrons. Therefore, the effect of low spray angles on the coating absorption capability and microstructure was investigated.

The planar Cu samples were sprayed with 90°, 70°, 50° and 30° spray angle. Al_2O_3/TiO_2 87/13 wt.% -22+5 μm powder was used for the experiments, in which the coating thickness for all the samples was constant and around 150 μm. Polished cross-sections of the samples and SEM images of the coating surface can be observed in Fig. 2.

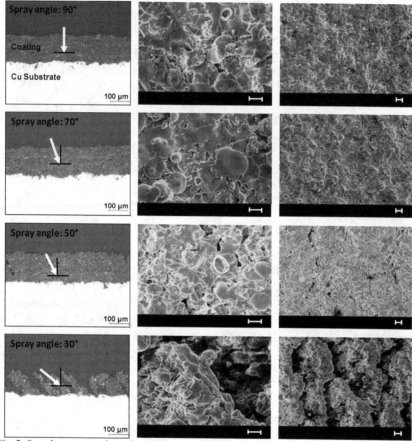

Fig. 2: Samples cross section micrographs and SEM pictures of the coatings surfaces deposited with spray angles between 90° and 30° [19].

It has been demonstrated that the spray angle has a strong influence on the deformation of the particles into splats and its morphology, varying as a consequence the properties of the coatings. The normal component of the in-flight particles velocity regarding the substrate surface is reduced with low spray angles, while the tangential velocity component, which controls the degree of splat elongation into elliptical shape, is increased [20]. Low spray angles lead to the formation of more elongated splats orientated in the direction of spraying [17], instead of the circular shaped splats obtained with normal angles [20]. The cross-section micrographs in Fig. 2 show the orientation of the lamellas in the direction of spraying, being the white regions attributed to TiO_2 phases, while the grey ones are Al_2O_3 [21].

Moreover, spray angles lower than 90°, which are associated with higher tangential velocities, result in a higher probability for the particles to rebound at impact with the substrate [20]. In this way, the deposition efficiency might be lower and maximum coating thickness is obtained for spraying operations perpendicular to the substrate surface [22]. Due to the deformation of the splats at impact on the substrate in the direction of spraying, the coated surface profile when spraying with off-normal angles presents a saw tooth shape [23], as it can be observed for the sample sprayed with 30°. Consequently, some regions of the saw tooth profile cannot be reached by the spray particles and remain uncovered, arising in those areas pores and increasing the coating roughness, as observed in the coated samples, see Fig. 3 (left).

The microwave absorption capability of the coatings sprayed with different angles was measured. As shown in Fig. 3 (right), the absorption of the coatings is similar for samples sprayed with angles between 90° and 50°, but decreases strongly when coating with 30°. This behaviour is related to the highly inhomogeneous coating microstructure characterized by high porosity and roughness obtained with low spray angles.

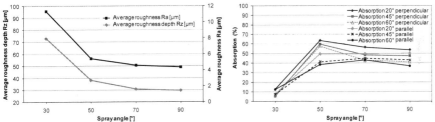

Fig. 3: Coating roughness (left) and microwave absorption (right) depending on the spray angle [19].

Coating material selection

In the research area of thermonuclear fusion, the use of Al_2O_3/TiO_2 to coat different components, mostly in the composition 87/13 wt.%, has been reported in the literature [3, 7-10]. With the application as absorbing microwave coatings, apart from Al_2O_3/TiO_2, the use of Cr_2O_3 and ZrO_2–$8Y_2O_3$ has been as well analysed by Spinicchia et al. [7]. Nanobashvili et al. [24] tested the microwave absorption behaviour of water stabilized plasma sprayed Al_2O_3, B_4C and Si coatings with different post treatments on steel substrates. In the present study, several spray materials were deposited on copper substrates and characterised in order to select the adequate one for the application of study. From pure Al_2O_3 to TiO_2, mixtures of Al_2O_3/TiO_2 with increasing content of TiO_2 were sprayed. Based on the literature, Cr_2O_3 and ZrO_2 were also investigated. All the used powders are fused and crushed and commercially available, with a grain size distribution of -20+5 μm. The coating thickness was 150 μm for all the materials. The results of absorption measurements for a frequency of 140 GHz are displayed in Fig. 4. Highest absorption rates were measured with the Al_2O_3/TiO_2 mixed powders with compositions between 87/13 wt.% and 50/50 wt.%. The pure oxide ceramic materials investigated showed worse performance.

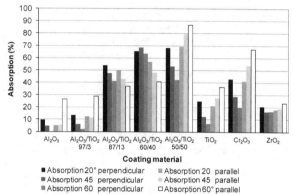

Fig. 4: Coating microwave absorption for different coating materials [19].

Al_2O_3/TiO_2 spray powders with composition 87/13 wt.% and 50/50 wt.% were selected for further analysis and system optimization. For each powder composition, two different grain sizes were used, -20+5 μm and -40+10 μm, in order to investigate its effect on the microwave absorption capability of the coatings. Coatings with a thickness from 50 to 250 μm approximately, were manufactured with each investigated spray material. Microwave absorption was measured. Results (see Fig. 5) show that the coating thickness is a determinant factor in the microwave absorption capability of the coatings. For each material, a thickness range can be identified in which the microwave absorption is maximised. In all the cases, except for the Al_2O_3/TiO_2 87/13 wt.% -20+5 μm coating, values higher than 90% were measured for the average absorption of all measurements at different angles of incidence and polarisations. The effect of the grain size distribution on the microwave absorption is not clear identified. For the materials with composition 87/13 wt.%, the thickness range in which maximised absorption is obtained is wider for the coatings sprayed with coarser powder than for those manufactured with the finer powder. This effect is less clearly visible for the 50/50 wt.% coatings.

Coating characterisation

The best absorbing coatings of each of the four powders analysed were characterised regarding their microstructure and mechanical properties (see Table I). Microhardness of Al_2O_3/TiO_2 coatings tends to decrease with the addition of TiO_2 [21], as can be seen from the results shown in table I. In this study the 50/50 wt.% coatings showed higher adhesion values than the samples coated with 87/13 wt.%, and for the same powder composition, higher adhesion was obtained for the samples sprayed with coarser powder. Moreover, the coarse grains lead to rougher coatings. No coating delamination or microcracks were detected in the samples. Infrared measurements were carried out on the samples and all the coatings presented emission coefficients over 0.9.

In order to analyse the phase changes that take place in the sprayed material during the coating process, XRD analysis of the four selected powders and the respective coatings were made. The X-Ray diffraction patterns of the powders are shown in Fig. 6 (left). Both powders with 13% TiO_2 content show as main phase α-Al_2O_3. The finer powder with particle size -20+5 μm presents also as minor phase stoichiometric TiO_2 in form of rutile. On the contrary, due to the higher content of TiO_2, Al_2O_3/TiO_2 50/50 wt.% spraying powders show a very different XRD pattern. Corundum is still present, but with lower intensity, and rutile can be as well be observed. Al_2TiO_5 with lattice parameters shifted in relation to those of the Al_2TiO_5 standard was detected.

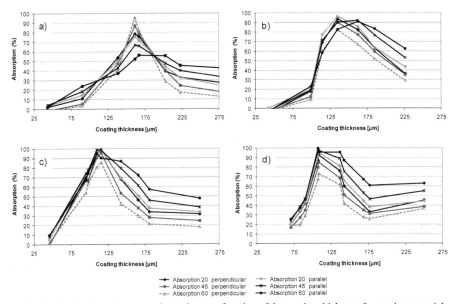

Fig. 5: Coating microwave absorption as a function of the coating thickness for coating material a) Al_2O_3/TiO_2 87/13 wt.%, -20+5 μm, b) Al_2O_3/TiO_2 87/13 wt.%, -40+10 μm, c) Al_2O_3/TiO_2 50/50 wt.%, -20+5 μm and d) Al_2O_3/TiO_2 50/50 wt.%, -40+10 μm.

Table I: Characterisation of the coatings with best microwave absorption for each Al_2O_3/TiO_2 spray powder [19]

Coating system	Coating thickness [μm]	Average microwave absorption (%)	Coating porosity (%)	Coating roughness Rz [μm]	Micro-hardness HV 0.1	Coating adhesion [MPa]
Al_2O_3/TiO_2 87/13, -20+5 μm	161.54	75.85	2.06	29.87	976.43	9.97
Al_2O_3/TiO_2 87/13, -40+10 μm	135.29	89.18	4.20	36.24	977.41	14.53
Al_2O_3/TiO_2 50/50, -20+5 μm	115.94	94.68	2.32	27.68	919.35	18.95
Al_2O_3/TiO_2 50/50, -40+10 μm	108.33	89.91	2.04	31.78	846.46	25.41

The crystalline composition of coatings sprayed with the four analysed powders, and with a thickness of approximately 150 μm determined by XRD can be seen in Fig. 6, right. The two coatings sprayed with Al_2O_3/TiO_2 87/13 wt.% show similar XRD patterns. Most of the corundum (α-Al_2O_3) found in the spraying powders has been transformed into the cubic γ-Al_2O_3, which is characteristic of APS coatings. This effect has been extensively discussed in the literature and was explained by McPherson attending to nucleation kinetics [25]. Due to its lower interfacial energy between crystal and liquid, γ-Al_2O_3 is more easily nucleated from the melt than α-Al_2O_3, being therefore the metastable phase, if cooled rapidly enough, the one retained to ambient temperature and therefore present in the coating [25]. The cooling rate of APS processes, which can be in the range of 10^6 K/s [26], strongly determines the phase composition of the coating. Generally, the higher the cooling rate, the more the γ phase is formed [27, 28]. However, some α-Al_2O_3 can be still found in the layers. The peaks corresponding to α-Al_2O_3 are slightly higher in intensity for the coarser powder and this fact may indicate a lower melting degree of the particles during spraying [29] in comparison with the finer powder. Rutile is also observed.

Figure 6.a and b (right) show the XRD patterns of the coatings sprayed with Al_2O_3/TiO_2 50/50 wt.%. The use of different grain sizes in the spraying powders gives rise to slight differences between the XRD results of the two coatings, which follow a very similar pattern. γ-Al_2O_3 is observed with lower intensity than in the 87/13 coatings and almost no corundum is found in the coatings. Rutile is as well observed. Al_2TiO_5 peaks are identified slightly shifted with respect to the standard peaks. This shift can be due to the formation and dissolution of Ti_3O_5 in Al_2TiO_5 under reducing conditions [30]. Comparing the Al_2TiO_5 peaks of coating a and b, the main peaks, which are between 20° and 50°, have higher intensity for the coating sprayed with finer powder. This effect could be explained attending to the fact that small powder particles have larger contact areas between each other and therefore, the particles might react stronger during spraying, forming Al_2TiO_5 [21].

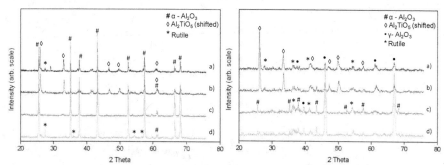

Fig. 6: XRD results of the analysed spray powders (left) and coatings (right): a) Al_2O_3/TiO_2 50/50 wt.%, -40+10 μm, b) Al_2O_3/TiO_2 50/50 wt.%, -20+5 μm, c) Al_2O_3/TiO_2 87/13 wt.%, -40+10 μm, d) Al_2O_3/TiO_2 87/13 wt.%, -20+5 μm.

Residual stresses in coating systems are generated during the production process due to the macroscopic and microscopic non-homogeneous elastic and elastic-plastic deformation after mechanical and/or thermal load [15]. Residual stresses arise as a result of temperature gradients, differences in thermophysical material properties and the cooling and solidification process and affect the structural properties of the coating, as well as its functionality and reliability [31]. The presence of critical residual stresses in the coating composite can lead to coating failure in form of delamination in the coating interface, crack network or plastic material deformation [15].

Residual stress measurements were carried out, as shown in Fig. 8 (left). The four coating systems show very low stresses. At the coating surface all the layers present tensile stresses, except the Al_2O_3/TiO_2 50/50 with powder size distribution -20+5 μm. During coating deposition, splats cool down after impact with the substrate and solidify transferring heat to the substrate and environment. This leads to contraction of the individual splats, which is constrained by the expansion of the substrate because of the rise in its temperature. As a result, quenching stresses of tensile nature arise in the coating [32]. When the substrate and coating temperatures are compensated, the composite cools down to room temperature. During this process the mismatch in CTE between the coating and substrate leads to the formation of thermal stresses, which in this case are of compressive nature due to the higher CTE of copper with a value of 20.3 10^{-6} 1/K [33], in comparison with the CTE of the ceramic coatings. However, the high thermal conductivity characteristic of copper of 401 W/K·m [33] may strongly reduce the thermal stresses. The final residual stresses situation in the coating results from the superposition of solidification and quenching and thermal stresses [32]. The use of a metallic bond coating to minimize the difference in CTE between coating and substrate was not necessary, as the residual stresses were low.

Al_2O_3/TiO_2 spray powders with compositions 87/13 wt.% and 50/50 wt.% showed good microwave absorption capability as well as mechanical stability. Both materials were selected with grain size distribution of -40+10 μm as the ones with highest potential for the application of study. As seen before, coarser grain sizes led in the case of 87/13 wt.% to a wider thickness range with high absorption capability. This is important for the coating of the water baffles, as the coating thickness might vary along the complex geometry of the component due to the different spray angles needed to perform the spraying process. Fig. 7 displays micrograph cross sections of the selected coatings.

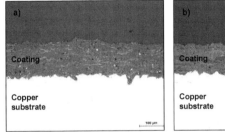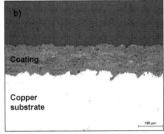

Fig. 7: Micrograph cross section of the selected coating systems to spray the water baffle prototypes: a) Al_2O_3/TiO_2 87/13 wt.%, -40+10 μm, b) Al_2O_3/TiO_2 50/50 wt.%, -40+10 μm.

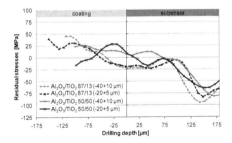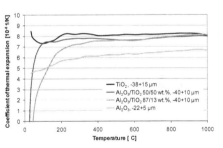

Fig. 8: Residual stress measurements of the best absorbing coatings of each analysed Al_2O_3/TiO_2 coating system (left) [19] and thermal expansion measurements of selected coatings (right).

For the two selected coating systems, dilatometry measurements were carried out. Thick coatings were sprayed with the two powders, removed from the substrate and cut to obtain samples of dimensions 3 x 4 x 6 mm^3 for the measurements. The results are displayed in Fig. 8 (right) with CTE measurements of Al_2O_3 and TiO_2 thermally sprayed coatings as reference. Coatings sprayed with

Al_2O_3/TiO_2 87/13 wt.% and 50/50 wt.% powders have similar CTE varying with the temperature between 7 - 8 10^{-6} 1/K. The strongest difference is seen at low temperatures. Similar results have been reported in the literature for Al_2O_3/TiO_2 coatings with different content in TiO_2 [34, 35].

Coating and testing of the water baffle prototypes
    In order to coat the water baffle prototypes, the robot trajectory had to be designed and the coating process optimised for planar substrates had to be adapted to the complex geometry of the water baffle. The mock-up was coated with a meander trajectory as shown in Fig. 9.

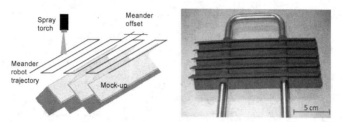

Fig. 9: Schematic view of the robot trajectory during coating (left) and coated mock-up (right) [19].

    The robot trajectory was designed in order to keep the coating thickness constant or as close as possible to the value which gave highest microwave absoprtion rates. With this aim, the meander offset as well as the number of coating cycles and combination of spray angles were varied and optimised.
    Although it was showed that spraying with angles lower than 30° strongly decreases the coating absorption, it was necessary to use off-normal angles in order to reach some areas of the water-baffle, see Fig. 10. The outer area of the chevrons was coated using a spray angle of 39°. The inner area was coated by two different spray operations with angles of 20° and 90°. Each spray angle was used from each side of the mock-up to achieve the complete coating of the component. Fig. 10 displays the areas of the substrate which are reached by each spray angle and cross section micrographs of the coatings obtained. The saw tooth profile, characteristic of low spray angles, can be observed in the coating sprayed with 20°. This inhomogeneous coating might reduce the overall absorption of the coating, obtaining therefore lower absorption values than on the planar substrates.
    Four baffles were coated, two of them with the optimised system of Al_2O_3/TiO_2 50/50 wt.% and two with 87/13 wt.%, all with powder grain size distribution of -40+10 μm. In-vessel components of W7-X, like the water baffle coated in this study, can be tested in an environment of isotropic 140 GHz radiation using the Microwave Stray Radiation Launch facility, MISTRAL [36]. The four coated prototypes and an uncoated one were tested in MISTRAL at the W7-X facilities. Results of the MISTRAL tests can be found in [19] and indicate a high absorption rate for the coated mock-ups compared to the uncoated one. The absorption of the mock-up coated with Al_2O_3/TiO_2 50/50 wt.% was of 60.3% and the baffle coated with Al_2O_3/TiO_2 87/13 wt.% showed absorption up to 75.7% [19].

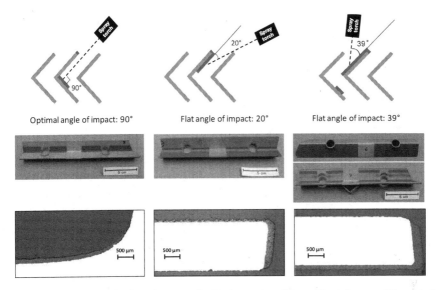

Fig. 10: Optimized combination of spray angles for the coating of the mock-up, images of the coated areas on the chevrons by each used angle and microscope cross sections of the coatings.

SUMMARY AND CONCLUSIONS

Microwave absorbing ceramic coatings applied on mock-ups were investigated with the aim of reducing the thermal load on the cryopanels in Wendelstein 7-X. Coatings were manufactured by Atmospheric Plasma Spray technology and characterised regarding their mechanical stability and their microwave absorption capability at a frequency of 140 GHz.

Al_2O_3/TiO_2 powders with different TiO_2 content and powder grain size distribution were sprayed and tested. It was found that coating thickness, grain size and composition (Al_2O_3-TiO_2 ratio) strongly influence the system absorption. For each powder composition, there are different coating thickness ranges at which the microwave absorption is maximized. Al_2O_3/TiO_2 powders with composition 50/50 and 87/13 wt.% and grain size distribution -40+10 μm were selected to coat the mock-ups.

The robot trajectory was optimized to coat the complete area of the mock-up, keeping the coating thickness at the appropriate value for high microwave absorption. The effect of the spray angle on the coating microstructure and functionality has been investigated. It was observed, that spray angles lower than 50° can strongly reduce the absorption behaviour of the coating. However, in order to coat the real components, it was necessary to use low spray angles. The mock-ups were sprayed using a combination of three spray angles.

Coated mock-ups were tested in MISTRAL at W7-X and showed absorption up to 75.7% in the case of Al_2O_3/TiO_2 87/13 wt.% coatings [19].

ACKNOWLEDGEMENTS

The authors would like to thank Dr. Walter Kasparek for the support with microwave absorption measurements.

REFERENCES

[1]W. M. Stacey: *Fusion. An Introduction to the Physics and Technology of Magnetic Confinement Fusion* (John Wiley & Sons Ltd, 1984).

[2]J. Ongena and G. VanOost, Energy for Future Centuries: Will Fusion be an Inexhaustible, Safe, Clean Energy Source?, *Trans. Fusion Technol.*, **37**, 3-15 (2000).

[3]J. Matejıcek, P. Chraska, and J. Linke, Thermal Spray Coatings for Fusion Applications—Review, *J. Therm. Spray Technol.*, **16 (1)**, 64-83 (2007).

[4]D. W. Sedgley, A. G. Tobin, T. H. Batzer, and W. R. Call, Cryopumping for Fusion Reactors, *Nucl. Eng. Des. Fus.*, **4**, 149-163 (1987).

[5]L.Wegener, Status of Wendelstein 7-X Construction, *Fusion Eng. Des.*, **84**, 106-112 (2009).

[6]S. Benhard, J. Boscary, H. Greuner, P. Grigull, J. Kißlinger, C. Li, B. Mendelevitch, T. Pirsch, N. Rust, S. Schweizer, A. Vorköper, and M. Weißgerber, Manufacturing of the Wendelstein 7-X divertor and wall protection, *Fusion Eng. Des.,* **75–79**, 463-468 (2005).

[7]N. Spinicchia, G. Angella, R. Benocci, A. Bruschi, A. Cremona, G. Gittini, A. Nardone, E. Signorelli, and E. Vassallo, Study of Plasma Sprayed Ceramic Coatings for High Power Density Microwave Loads, *Surf. Coat. Technol.*, **200**, 1151-1154 (2005).

[8]A. Bruschi, S. Cirant, F. Gandini, G. Granucci, V. Mellera, V. Muzzini, A. Nardone, A. Simonetto, C. Sozzi, and N. Spinicchia, Design of a high-power load for millimetre-wave Gaussian beams, *Nucl. Fusion*, **43 (11)**, 1513-1519 (2003).

[9]F. Brossa, G. Rigon, and B. Looman, Behaviour of Plasma Spray Coatings under Disruption Simulation, *J. Nucl. Mater.*, **155-157**, 267-272 (1988).

[10]K. Likin, A. Fernández, and R. Martín, in: Proceedings of the 26th International Conference on Infrared and Milimeter Waves, COREP, Toulouse, 4–88 (2001).

[11]L. Pawlowski: *The Science and Engineering of Thermal Spray Coatings, 2nd ed.* (John Wiley & Sons Ltd, Chichester, England, 2008).

[12]P. Fauchais, A. Vardelle, and B. Dussoubs, Quo Vadis Thermal Spraying?, *J. Therm. Spray Technol.*, **10**, 44-66 (2001).

[13]P. Fauchais, Understanding plasma spraying: Topical review, *J. Phys. D. Appl. Phys.*, **37**, 86-108 (2004).

[14]H. Du, J. H. Shin, and S. W. Lee, Study on Porosity of Plasma-Sprayed Coatings by Digital Image Analysis Method, *J. Therm. Spray Technol.*, **14**, 453-461 (2005).

[15]R. Gadow, M.J. Riegert-Escribano, and M. Buchmann, Residual stress analysis in thermally sprayed layer composites, using the hole milling and drilling method, *J. Therm. Spray Technol.*, **14**, 100-108 (2004).

[16]R. Wacker, F. Leuterer, D. Wagner, H. Hailer, and W. Kasparek, Characterization of absorber materials for high-power millimetre waves, in: Proceedings of the 27th International Conference on Infrared and milimeter Waves, IEEE Press, New York, 159-160 (2002).

[17]G. Montavon, S. Sampath, C.C. Berndt, H. Herman, and C. Coddet, Effect of the spray angle on splat morphology during thermal spraying, *Surf. Coat. Technol.*, **91**, 107-115 (1997).

[18]J. Ilavsky, A. Allen, G. Long, and S. Krueger, Influence of Spray Angle on the Pore and Crack Microstructure of Plasma-Sprayed Deposits, *J. Am. Ceram. Soc.*, **80 (3)**, 733-742 (1997).

[19]M. Floristán, P. Müller, A. Gebhardt, A. Killinger, R. Gadow, A. Cardella, C. Li, R. Stadler, G. Zangl, M. Hirsch, H. P. Laqua, and W. Kasparek, Development and Testing of 140 GHz Absorber Coatings for the Water Baffle of W7-X Cryopumps, *Fusion Eng. Des.* (2010), doi:10.1016/j.fusengdes.2010.12.015.

[20]C. W. Kang, H. W. Ng, and S. C. M. Yu, Imaging Diagnostics Study on Obliquely Impacting Plasma-Sprayed Particles Near to the Substrate, *J. Therm. Spray Technol.*, **15 (1)**, 118-130 (2006).

[21]R. Yilmaz, A. O. Kurt, A. Demir, and Z. Tatli, Effects of TiO_2 on the mechanical properties of the Al_2O_3–TiO_2 plasma sprayed coating, *J. Eur. Ceram. Soc.,* **27**, 1319-1323 (2007).

[22]W. Tillmann, E. Vogli, and B. Krebs, Influence of the spray angle on characteristics for atmospheric plasma sprayed hard material based coatings, *J. Therm. Spray Technol.,* **17 (5-6)**, 948-955 (2008).

[23]M. F. Smith, R.A. Neiser, and R.C. Dykhuizen, in: Proceeding of the 7th National Thermal Spray Conference, C.C. Berndt., ASM International, Materials Park, Ohio, 603 (1995).

[24]S. Nanobashvili, J. Matejícek, F. Zácek, J. Stockel, P. Chráska, and V. Brozek, Plasma Sprayed Coatings for RF Wave Absorption, *J. Nucl. Mater.,* **307-311**, 1334-1338 (2002).

[25]R. McPherson, Formation of Metastable Phases in Flame and Plasma-Preapared Alumina, *J. Mater. Sci.,* **8 (6)**, 851-858 (1973).

[26]L. Zhao, K. Seemann, A. Fischer, and E.Lugscheider, Study on Atmospheric Plasma Spraying of Al_2O_3 Using On-Line Particle Monitoring, *Surf. Coat. Technol.,* **168**, 186-190 (2003).

[27]P. Chráska, J. Dubsky, K. Neufuss and J. Píacka, Alumina-Base Plasma-Sprayed Materials Part I: Phase Stability of Alumina and Alumina-Chromia, *J. Therm. Spray Technol.,* **6 (3)**, 320-325 (1997).

[28]L. Pawlowsky, The Relationship Between Structure and Dielectric Properties in Plasma-Sprayed Alumina Coatings, *Surf. Coat. Technol.,* **35**, 285-298 (1988).

[29]S. Yilmaz, An Evaluation of Plasma-Sprayed Coatings Based on Al_2O_3 and Al_2O_3 13 wt% TiO_2 with Bond Coat on Pure Titanium Substrate, *Ceram. Int. Vol.,* **35**, 2017-2022 (2009).

[30]F.-L. Toma, C.C. Stahr, L.-M. Berger, S. Saaro, M. Herrmann, D. Deska, and G. Michael, Corrosion Resistance of APS- and HVOF Sprayed Coatings in the Al_2O_3-TiO_2 System, *J. Therm. Spray Technol.,* **19 (1, 2)**, 137-147 (2010).

[31]M. Wenzelburger, D. López, and R. Gadow, Methods and application of residual stress analysis on thermally sprayed coatings and layer composites, *Surf. Coat. Technol.,* **201**, 1995-2001 (2006).

[32]M. Wenzelburger, R. Gadow, and M. Escribano, Modeling of thermally sprayed coatings on light metal substrates: - layer growth and residual stress formation, *Surf. Coat. Technol.,* **180-181**, 429-435 (2004).

[33]L. Xie, D. Chen, E. H. Jordan, A. Ozturk, F. Wu, X. Ma, B. M. Cetegen, and M. Gell, Formation of vertical cracks in solution-precursor plasma-sprayed thermal barrier coatings, *Surf. Coat. Technol.,* **201**, 1058-1064 (2006).

[34]S. Rangaswamy and H. Herman, Thermal expansion Study of Plasma-Sprayed Oxide Coatings, *Thin Solid Films,* **73**, 43-52 (1980).

[35]C. J. Friedrich: *Atmosphärisch plasmagespritzte dielektrische Oxidschichten für Ozongeneratoren* (Shaker Verlag, Aachen, Germany, 2002).

[36]S. Ullrich, Charakterisierung einer Materialtestkammer als Millimeterwellen-Resonator bei 140 GHz, Diploma Thesis, University of Greifswald, 2005 (summary in Stellarator News 98 (May 2005)) available on the web: http://www.ornl.gov/ sci/fed/stelnews/.

Thermal Barrier Coatings

HOT CORROSION OF POTENTIAL THERMAL BARRIER COATING MATERIAL $(SM_{1-x}YB_x)_2ZR_2O_7$ BY V_2O_5 AND NA_2SO_4

Jia-Hu Ouyang , Sa Li, Zhan-Guo Liu, Yu Zhou

Institute for Advanced Ceramics, Department of Materials Science, Harbin Institute of Technology, 92 West Da-Zhi Street, Harbin 150001, China

ABSTRACT

$(Sm_{1-x}Yb_x)_2Zr_2O_7$ ($x = 0$, 0.5, 1.0) powders are prepared by chemical-coprecipitation and calcination method, and then pressureless-sintered at 1700 °C for 10 h. $Sm_2Zr_2O_7$ has a pyrochlore-type structure, while $SmYbZr_2O_7$ and $Yb_2Zr_2O_7$ have a defect fluorite-type structure. Thermal expansion coefficient and thermal diffusivity of $(Sm_{1-x}Yb_x)_2Zr_2O_7$ are studied by a high-temperature dilatometer and a laser flash diffusivity technique from room temperature to 1400 °C. Hot corrosion tests between $(Sm_{1-x}Yb_x)_2Zr_2O_7$ and three corrosive agents including V_2O_5, Na_2SO_4, and a $V_2O_5+Na_2SO_4$ mixture, are carried out from 600 to 1100 °C for 2 h and 8 h in air, respectively. Different reaction products of ZrV_2O_7, $LnVO_4$ and m-ZrO_2 are identified depending upon the hot corrosion conditions, for example, ZrV_2O_7 and corresponding $LnVO_4$ at 600 °C for 2 h and 8 h, namely $SmVO_4$, $(Sm,Yb)VO_4$, $YbVO_4$, respectively; ZrV_2O_7, m-ZrO_2 and $LnVO_4$ at 700 °C for 2 h; m-ZrO_2 and $LnVO_4$ either at 800~1100 °C for 2 h or at 700~1100 °C for 8 h. No reaction products are identified on Na_2SO_4-coated $(Sm_{1-x}Yb_x)_2Zr_2O_7$ at 900~1100 °C. However, m-ZrO_2 and corresponding $LnVO_4$ are found after $(Sm_{1-x}Yb_x)_2Zr_2O_7$ exposed to $Na_2SO_4+V_2O_5$ (mole ratio = 1:1) at temperatures of 600~1100 °C. Those results are explained based on phase diagram theory, and the principles for crystal growth are used to illustrate the morphologies of reaction products $LnVO_4$.

1. INTRODUCTION

Rare-earth zirconate ceramics used as thermal barrier coatings (TBCs) have attracted increasing interest in recent years due to their distinctly lower thermal conductivity than the common TBC material, 4–4.5 mol.% Y_2O_3–ZrO_2 (YSZ).[1-6] The order-disorder transition temperature of rare-earth zirconates is clearly higher than that (1200 °C) of tetragonal to monoclinic phase transformation in YSZ. $Sm_2Zr_2O_7$ exhibits a transition from ordered pyrochlore structure to disordered fluorite structure at 2000 °C, while $Yb_2Zr_2O_7$ is phase stable up to its melting point. As a result, rare-earth zirconate ceramics are potential candidates for high-temperature TBCs applications.

In practice, TBCs may encounter various corrosive environments. When TBCs are working in low-quality fuels containing impurities such as sodium, sulfur, phosphorus or especially vanadium, the hot corrosion attack of molten salt will accelerate its degradation and reduce its lifetime severely.[7-11] Among those contaminations, chemical interactions between vanadium pentoxide and zirconia-based coatings are fastest and therefore the most deleterious.[9] As a result, it is of great significance to explore a comprehensive understanding on hot corrosion mechanisms of rare-earth zirconates against V_2O_5. Marple et al. compared the hot corrosion behavior of $La_2Zr_2O_7$ and YSZ coatings to vanadium-containing compounds at 900 and 1000 °C.[11] $La_2Zr_2O_7$ coatings were relatively resistant to attack by V_2O_5. $La_2Zr_2O_7$ coatings remained well bonded to the substrate following exposure to V_2O_5, and contained only minor amounts of $LaVO_4$. In contrast, increased microcracking and spallation of

---

E-mail address: ouyangjh@hit.edu.cn (J.-H. Ouyang).

the 8YSZ coatings were observed, accompanied by the formation of YVO_4 and monoclinic ZrO_2. Xu *et al.* reported the hot corrosion performances of YSZ, $La_2Zr_2O_7$ and $La_2(Zr_{0.7}Ce_{0.3})_2O_7$ coatings in the presence of molten mixture of $Na_2SO_4 + V_2O_5$ at 900 °C for 100 h.[12] The reaction between $NaVO_3$ and Y_2O_3 produced YVO_4 and caused the progressive destabilization transformation of YSZ. Sodium vanadate and sodium sulfate also resulted in rapid disintegration of the $La_2Zr_2O_7$ coating. However, the $La_2(Zr_{0.7}Ce_{0.3})_2O_7$ coatings remained perfect without significant degradation. $Gd_2Zr_2O_7$ synthesized by solid-state reaction had two different corrosion mechanisms when reacted with molten V_2O_5 in a temperature range from 700 to 850 °C, which was explained based on the thermal instability of ZrV_2O_7.[13] To date, only a very limited amount of experiments on hot corrosion of rare-earth zirconates against vanadium pentoxide have been done in the open literatures; however, hot corrosion mechanisms on reactions between $(Sm_{1-x}Yb_x)_2Zr_2O_7$ and V_2O_5 still remain unclear.

In this work, hot corrosion behavior of $(Sm_{1-x}Yb_x)_2Zr_2O_7$ ($x = 0, 0.5, 1.0$) ceramics were investigated by exposing them to different corrosive agents including V_2O_5, Na_2SO_4 and a mixture of $Na_2SO_4 + V_2O_5$ under various heat-treatment conditions. Special attention was paid to the phase constituents as well as morphological characteristics of hot-corroded surfaces. The hot corrosion mechanisms of $(Sm_{1-x}Yb_x)_2Zr_2O_7$ ceramics against different corrosive agents were investigated.

2. EXPERIMENTAL

$(Sm_{1-x}Yb_x)_2Zr_2O_7$ ($x = 0, 0.5, 1.0$) ceramic powders were prepared through chemical- coprecipitation and calcination method with starting powders of samarium oxide, ytterbium oxide (Rare-Chem Hi-Tech Co., Ltd., Beijing, China; purity $\geq$ 99.99%) and zirconium oxychloride (Zibo Huantuo Chemical Co. Ltd., Huizhou, China; Analytical). Details of the powder preparation process can be found in our previous work.[14] The $(Sm_{1-x}Yb_x)_2Zr_2O_7$ ($x = 0, 0.5, 1.0$) powders were molded by uniaxial stress, and the molded samples were further compacted by cold isostatic pressing at 280 MPa for 5 min. The compacts were pressureless-sintered at 1700 °C for 10 h at a heating rate of 5 °C/min in air. From X-ray diffraction (XRD) measurements (Fig. 1), $Sm_2Zr_2O_7$ exhibits a pyrochlore-type structure, while both $SmYbZr_2O_7$ and $Yb_2Zr_2O_7$ have a defect fluorite-type structure.[15] The relative density of $Sm_2Zr_2O_7$, $Yb_2Zr_2O_7$ and $SmYbZr_2O_7$ ceramics was measured to be 97.6%, 96.0% and 96.6% using the Archimedes principle, respectively.

The thermal diffusivity was measured using the laser flash diffusivity technique (Netzsch LFA 427, Germany) from room temperature to 1400 °C in an argon atmosphere. The disk-shaped samples were 12.7 mm in diameter and 1.5 mm in thickness. Prior to the thermal diffusivity measurement, the specimen surfaces were coated with a thin layer of sprayed colloidal graphite. Thermal diffusivity measurement of the specimen was carried out three times at each temperature. The specific heat capacities, as a function of temperature, were determined from the chemical compositions of $(Sm_{1-x}Yb_x)_2Zr_2O_7$ ceramics and the heat-capacity data of the constituent oxides (Sm_2O_3, Yb_2O_3 and ZrO_2) obtained from the literature,[16] in conjunction with the Neumann-Kopp rule.[17] The thermal conductivity k was determined from the heat capacity C_p, density ρ and thermal diffusivity λ, and using the equation

$$k = C_p \lambda \rho \tag{1}$$

The high-temperature densities of all specimens were calibrated by the dilatometric measurement data. Because the specimens were not fully dense, the measured thermal conductivity data were corrected for the residual porosity of the samples.[3]

The linear thermal expansion coefficients of sintered ceramics were determined with a high-temperature dilatometer (Netzsch DIL 402C, Germany) from room temperature to 1400°C in an argon atmosphere. Data were continuously recorded at a heating rate of 5 °C/min during heating, and they were corrected using the known thermal expansion coefficient of a certified standard alumina. The specimens have the dimensions of about 4 mm×4 mm×20 mm.

The specimens for hot corrosion tests were machined to the size of 10 mm × 10 mm × 3 mm from the as-sintered $(Sm_{1-x}Yb_x)_2Zr_2O_7$ ceramics. The specimens were ground to 1500 grit finish, ultrasonically degreased in acetone, and dried at 100 °C over night. The corrosive agent powders were spread uniformly over the surfaces of $(Sm_{1-x}Yb_x)_2Zr_2O_7$ specimens at a concentration of 20 mg/cm^2 using a very fine glass rod that was ultrasonically cleaned and dried. The as-prepared $(Sm_{1-x}Yb_x)_2Zr_2O_7$ specimens were then placed in a zirconia crucible, which was subsequently covered with a thin zirconia sheet during heat treatment. The V_2O_5-coated $(Sm_{1-x}Yb_x)_2Zr_2O_7$ specimens were isothermally heat-treated at temperatures of 600–1100 °C for 2 h and 8 h; Na_2SO_4-coated $(Sm_{1-x}Yb_x)_2Zr_2O_7$ was isothermally heat-treated in the temperature range of 900–1100 °C for 2 h and 8 h; meanwhile, the $(Na_2SO_4 + V_2O_5)$-coated $(Sm_{1-x}Yb_x)_2Zr_2O_7$ specimens were isothermally heat-treated at temperatures of 600–1100 °C for 2 h and 8 h.

After hot corrosion tests, crystal structures of hot corroded specimens were identified by an X-ray diffractometer (Rigaku D/Max-2200VPC, Tokyo, Japan) with Cu $K\alpha$ radiation at a scan rate of 3 °/min. The microstructural analysis of hot corroded specimens was carried out with a scanning electron microscope (FEI Quanta 200F, Eindhoven, the Netherlands) equipped with energy-dispersive X-ray spectroscopy operating at 30 kV. A thin carbon coating was sputtered onto the surface of hot corroded specimens to ensure good electrical conductivity.

3. RESULTS
3.1 Thermal conductivity and thermal expansion of $(Sm_{1-x}Yb_x)_2Zr_2O_7$ ceramics
Thermal conductivity of $(Sm_{1-x}Yb_x)_2Zr_2O_7$ ceramics as a function of temperature is shown in Fig. 2. The error bars are omitted as they are smaller than the symbols. Clearly, the measured thermal conductivity of $(Sm_{1-x}Yb_x)_2Zr_2O_7$ ceramics decreases gradually with increasing temperature from room temperature to 800 °C, which is attributed to the lattice thermal conduction. However, thermal conductivity of $(Sm_{1-x}Yb_x)_2Zr_2O_7$ ceramics slightly increases with further increasing temperature up to 1400 °C, which may be attributed to the increased radiation contribution with increasing temperature, also known as photon thermal conductivity. The thermal conductivity of $SmYbZr_2O_7$ ceramic is lower than that of $Sm_2Zr_2O_7$ or $Yb_2Zr_2O_7$ ceramic in this study. Thermal conductivity of $(Sm_{1-x}Yb_x)_2Zr_2O_7$ ceramics in this investigation is located within the range of 1.40–1.99 W·m^{-1}·K^{-1} from room temperature to 1400 °C, which indicates that $(Sm_{1-x}Yb_x)_2Zr_2O_7$ ceramics are potential candidates for high temperature TBCs applications.

Thermal expansion coefficients of $(Sm_{1-x}Yb_x)_2Zr_2O_7$ ceramics are shown in Fig. 3. Thermal expansion coefficients of $(Sm_{1-x}Yb_x)_2Zr_2O_7$ ceramics increase rapidly from room temperature to about 300 °C, which is caused by the nonlinear increase of instrument temperature. Thermal expansion coefficients of $(Sm_{1-x}Yb_x)_2Zr_2O_7$ ceramics increase with increasing temperature, which is a typical characteristic of solid materials as the atomic spacing increases with the increase of temperature. With increasing Yb content, thermal expansion coefficients of $(Sm_{1-x}Yb_x)_2Zr_2O_7$ ceramics gradually decrease at identical temperature levels. At 1250 °C, thermal expansion coefficients of $(Sm_{1-x}Yb_x)_2Zr_2O_7$ ceramics are within the range of 10.6–11.9 × 10^{-6} K^{-1}.

3.2 Hot corrosion behavior of V$_2$O$_5$-coated (Sm$_{1-x}$Yb$_x$)$_2$Zr$_2$O$_7$ samples

Vanadium in low-quality fuels is oxidized to form V$_2$O$_5$ in high operating temperature environment. The melting point of V$_2$O$_5$ is 690 °C. In this study, hot corrosion tests of V$_2$O$_5$-coated (Sm$_{1-x}$Yb$_x$)$_2$Zr$_2$O$_7$ samples at different temperatures of 600–1100 °C for different holding time of 2–8 h were conducted to fully exploit the hot corrosion mechanisms of (Sm$_{1-x}$Yb$_x$)$_2$Zr$_2$O$_7$ ceramics against molten V$_2$O$_5$. Figure 4 shows XRD patterns obtained from the V$_2$O$_5$-coated Sm$_2$Zr$_2$O$_7$ samples heat-treated at temperatures of 600–1100 °C in air. After hot corrosion tests at 600 °C for 2 h in air, the newly evolved peaks on the hot corroded surface include zirconium vanadate (ZrV$_2$O$_7$, JCPDS no. 16–0422) and samarium vanadate (SmVO$_4$, JCPDS no.72–0279). In order to investigate the influence of holding time on the reaction products, some V$_2$O$_5$-coated Sm$_2$Zr$_2$O$_7$ samples were isothermally heat treated at 600 °C for a relatively long time of 8h in an electric furnace. From Fig. 4, the phase constituents of the corroded surface after thermal exposure for 8h are quite similar to those obtained at 600 °C for 2 h. However, the hot corrosion products at 700 °C for 2 h in air are mainly composed of SmVO$_4$ and monoclinic zirconia (m-ZrO$_2$, JCPDS no.37–1484), together with small amounts of ZrV$_2$O$_7$. When the holding time increases from 2 h to 8 h at 700 °C, no ZrV$_2$O$_7$ peaks are identified from the XRD spectrum on the corroded surface after heat treatment except SmVO$_4$ and m-ZrO$_2$. The XRD results after hot corrosion tests at 800 °C or above for 2 h and 8 h are similar to those obtained at 700 °C for 8 h, all of which demonstrate strong diffraction peaks of SmVO$_4$ and weak peaks of m-ZrO$_2$. As revealed in Figs. 5 and 6, for the V$_2$O$_5$-coated Yb$_2$Zr$_2$O$_7$ and SmYbZr$_2$O$_7$ specimens heat-treated at 600–1100 °C for different holding time of 2–8 h, the phase constituents after V$_2$O$_5$-induced hot corrosion are similar to that of V$_2$O$_5$-coated Sm$_2$Zr$_2$O$_7$ samples. Different reaction products of zirconium vanadate (ZrV$_2$O$_7$, JCPDS no. 16–0422), monoclinic zirconia (m-ZrO$_2$, JCPDS no.37–1484) and LnVO$_4$ are identified depending upon the hot corrosion conditions. After hot corrosion tests at 600 °C for 2 h and 8 h, the hot corroded products contain ZrV$_2$O$_7$ and the corresponding rare-earth orthovanadates of LnVO$_4$, namely ytterbium vanadate (YbVO$_4$, JCPDS no.72–0271) and (Sm,Yb)VO$_4$ solid solution. ZrV$_2$O$_7$, m-ZrO$_2$ and LnVO$_4$ are formed on the hot corroded surfaces at 700 °C for 2 h. However, m-ZrO$_2$ and LnVO$_4$ are found after hot corrosion tests either at 800–1100 °C for 2 h or at 700–1100 °C for 8 h. The XRD results of V$_2$O$_5$-coated Yb$_2$Zr$_2$O$_7$ and SmYbZr$_2$O$_7$ samples heat-treated at temperatures of 600–1100 °C for different holding time are summarized in Tables 1 and 2. As can be seen, the substrate peaks are clearly detected after hot corrosion tests, which indicate that the thickness of corrosion scales on Yb$_2$Zr$_2$O$_7$ and SmYbZr$_2$O$_7$ substrates is quite small. Figure 5 shows backscattered electron images of the cross-sections of the corroded layers for V$_2$O$_5$-coated (Sm$_{1-x}$Yb$_x$)$_2$Zr$_2$O$_7$ specimens heat-treated at 800 °C for 8 h in air.

Surface morphology of (Sm$_{1-x}$Yb$_x$)$_2$Zr$_2$O$_7$ ceramics hot-corroded at 600–1100 °C for different holding time of 2–8 h exhibits similar characteristics, and therefore here we only present the typical surface morphology of V$_2$O$_5$-coated Sm$_2$Zr$_2$O$_7$, as shown in Fig. 6. SEM micrograph of the V$_2$O$_5$-coated Sm$_2$Zr$_2$O$_7$ samples heat-treated at 600 °C for 2 h is shown in Fig. 6(a). Two different morphologies of hot corroded products, marked as A and B, respectively, are observed. Product A exhibits a cubic-shape, while B is irregularly-shaped. In combination with the above XRD results, EDS analyses (Fig. 7(a) and (b)) demonstrate that A is ZrV$_2$O$_7$ and B is SmVO$_4$. The surface morphology of the V$_2$O$_5$-coated Sm$_2$Zr$_2$O$_7$ heat-treated at 600 °C for 8 h exhibits the similar characteristics as that of the hot-corroded for 2 h at 600 °C, and therefore it is not shown here. SEM micrograph obtained on the surface of Sm$_2$Zr$_2$O$_7$ specimen after thermal exposure to molten V$_2$O$_5$ at 800 °C for 2 h, is shown in Fig. 6(b).

From the EDS results (Fig. 7(c) and (d)) obtained at different regions of C and D, the elements identified are consistent with the presence of reaction products of m-ZrO_2 (region C) and $SmVO_4$ (region D). It can be seen that the morphologies of $SmVO_4$ in Fig. 6 (a and b) are quite similar, while the morphologies of ZrV_2O_7 in Fig. 6(a) and m-ZrO_2 in Fig. 6(b) exhibit a little difference in the size of reaction products. The size of ZrV_2O_7 in Fig. 6(a) is about 5–10 μm, which is larger than that of m-ZrO_2 (about 2–5 μm) in Fig. 6(b). Figure 6(c) is a representative SEM image showing the surface morphology of the hot corroded samples at 700 °C for 2 h. The EDS spectra (not shown here) obtained at different regions of E, F and G in Fig. 6(c) confirmed the presence of elements in good agreement with the formation of reaction products of ZrV_2O_7 (region E), $SmVO_4$ (region F) and m-ZrO_2 (region G). For the sake of brevity, the microstructures of the specimens hot corroded at 700–1100 °C for 8 h and 900–1100 °C for 2 h are not shown here because of their similarities to that of 800 °C for 2 h.

3.3 Hot corrosion behaviour of Na_2SO_4-coated $(Sm_{1-x}Yb_x)_2Zr_2O_7$ samples

XRD analysis was performed on the corrosion samples after exposure to Na_2SO_4 molten salt for 8 h at 900–1100 °C, as shown in Fig. 8. The dominant phase after hot corrosion was $(Sm_{1-x}Yb_x)_2Zr_2O_7$ with a small fraction of Na_2SO_4. Since only original phases were detected and no newly evolved peaks appeared, it is concluded that no evident chemical interaction occurs in this situation, which is also confirmed by the backscattered micrograph of the hot-corroded samples (Fig. 9). Figure 9 shows backscattered electron images of surface morphologies of Na_2SO_4-coated $(Sm_{1-x}Yb_x)_2Zr_2O_7$ specimen heat-treated at 900 °C for 8 h in air. Na_2SO_4 is detected at the positions of the pores on the substrate. Since molten Na_2SO_4 had better mobility at temperatures of 900–1100 °C, all the molten Na_2SO_4 would penetrate into the zirconate substrates if given sufficient corrosion time, i.e. 8 h.

3.4 Corrosion behavior of Na_2SO_4+V_2O_5-coated $(Sm_{1-x}Yb_x)_2Zr_2O_7$ sample

Vanadium compounds combine with sodium sulfate to form a eutectic liquid with a lower melting point, thus expanding the temperature range of hot corrosion. Fig. 10 shows XRD patterns obtained from Na_2SO_4+V_2O_5-coated $Sm_2Zr_2O_7$ specimen heat-treated for 8h at temperatures of 900–1100 °C. It demonstrates sharp diffraction peaks of samarium vanadate ($SmVO_4$, JCPDS no.72–0279) and weak peaks of monoclinic zirconia (m-ZrO_2, JCPDS no.37–1484), which originate from the corrosion products.

Typical surface morphology of the Na_2SO_4+V_2O_5-coated $Sm_2Zr_2O_7$ specimen heat-treated at 700–1100 °C for 2 h is shown in Fig. 11. There are two different morphologies of corrosion products are observed. EDS analysis (not shown here) was made to identify the compositions of the rod-like and the granular corrosion products. In combination with the above XRD results, the rod-like product is $SmVO_4$ and the granular product is m-ZrO_2. With increasing hot corrosion temperature, the oriented growth of $SmVO_4$ becomes obvious and the length of $SmVO_4$ rod prolongs. For the sake of brevity, surface morphology of $SmYbZr_2O_7$ and $Yb_2Zr_2O_7$ ceramics hot-corroded at 600–1100 °C for different holding time of 2–8 h is not provided in this paper.

4 DISCUSSION

4.1 Hot corrosion mechanisms in V_2O_5 melts

Vanadium pentoxide-coated $Sm_2Zr_2O_7$ specimen heat-treated at 800 °C for 8 h is observed to be degraded severely. The hot corrosion mechanism can be proposed based on the phase diagrams. The V_2O_5–Sm_2O_3 and V_2O_5–ZrO_2 binary phase diagrams were reported in literatures.[18,19] However, no phase equilibrium diagram in V_2O_5–$Sm_2Zr_2O_7$ system is established up to now. The chemical interaction between V_2O_5 and $Sm_2Zr_2O_7$ could be evaluated through a combination of V_2O_5–Sm_2O_3 and V_2O_5–ZrO_2 binary phase diagrams, considering that $Sm_2Zr_2O_7$ can be viewed as a compound which is formed by the introduction of Sm_2O_3 to cubic ZrO_2 oxides at a ZrO_2/Sm_2O_3 mole ratio of 2:1.[20] Thin film of fused salts can work as acid/base solvents to dissolve ceramic oxides and induce hot corrosion of ceramic oxides.[21] The $Sm_2Zr_2O_7$ substrate is locally dissolved by molten V_2O_5 and the local equilibrium is established. The whole system might be far from equilibrium, but the local equilibrium is believed to be maintained at the solid-liquid interface, for the atomic transfer across the solid-liquid interface is quite fast. As the reaction proceeds, the variations in chemical composition of the liquid containing V, Sm, Zr and O occur, and $SmVO_4$ as well as ZrV_2O_7 precipitates out of the liquid once they exceed the corresponding temperature-dependent saturation limits. Therefore, hot corrosion products can be formed at the molten salt-substrate interfaces through a dissolution-precipitation process[22]: 1) the liquid-solid system firstly reaches local equilibrium after dissolution of ceramic oxides, which means that the solid phases are locally in an equilibrium state with the liquid phase surrounding them; 2) depending on its local composition, the liquid dissolves the solid phases that are not in equilibrium with it and precipitates new phase after saturated; 3) mass transport and chemical reactions take place during the hot corrosion, which produce concentration gradients. Figure 11 presents the hot corrosion mechanism of $Sm_2Zr_2O_7$ against V_2O_5 and illustrates the formation process of hot corrosion products. It is worthy mentioning that the reactions between vanadium compounds and ceramic oxides are found to follow a Lewis acid-base mechanism, where the acid vanadium compounds will react more readily with the ceramic oxides that have a stronger basicity.[23] As is reported, the basicity of samarium oxide, ytterbium oxide and zirconium dioxide follows such order: $Sm_2O_3 > Yb_2O_3 > ZrO_2$, indicating that molten V_2O_5 has a larger tendency to react with Sm_2O_3 than ZrO_2.[24] According to the V_2O_5–Sm_2O_3 binary phase diagram,[18,24] $SmVO_4$, which is stable up to 1440 °C, can be obtained at 800 °C. The peaks of $SmVO_4$ are observed in the XRD pattern of corrosion products (Fig. 4). In the system of V_2O_5–ZrO_2,[19] ZrV_2O_7 is produced as an intermediate compound and it incongruently melts to be a mixture of m-ZrO_2 and a liquid containing V, Zr and O at 747 °C. The molten V_2O_5 generated from incongruent melting continues to react with ZrO_2 until it is depleted. Therefore, $SmVO_4$ and m-ZrO_2 are observed as the final reaction products in the experiment. The detailed mechanism of such a process was reported in previously published work, in which the corrosion product of ZrV_2O_7 was found after hot corrosion reaction of V_2O_5-coated $Yb_2Zr_2O_7$ at temperatures of 600 and 700 °C.[25] ZrV_2O_7 was made through a solid-state reaction of ZrO_2 and V_2O_5 by Buchanan et al., who believed that the process was very slow.[26] However, such conclusion is not applicable in this investigation, because the reaction is controlled by the diffusion of Y atoms in Buchanan's experiment, which is slow at 720 °C. On the contrary, it is believed that the kinetics of $SmVO_4$ development at 800 °C are dominated by the formation of the eutectic liquid between V_2O_5 and Sm_2O_3, which would afford rapid diffusion of Sm. The reaction of V_2O_5 and YSZ was also studied by Mohan et al., who found that ZrV_2O_7 appeared at 720 °C and disappeared at 800 °C.[27] He attributed the two different reactions observed to the thermal instability of ZrV_2O_7. Based on the analysis above, the hot corrosion mechanism of V_2O_5-coated $Sm_2Zr_2O_7$ samples heat-treated at 800 °C for 8 h can be expressed as:

$$V_2O_5(l) + Sm_2Zr_2O_7(s) \rightarrow 2SmVO_4(s) + 2m\text{-}ZrO_2(s) \qquad (2)$$

Hot corrosion mechanisms of $SmYbZr_2O_7$ and $Yb_2Zr_2O_7$ are similar to that of $Sm_2Zr_2O_7$, therefore it is reasonable to conclude that the chemical interaction of $SmYbZr_2O_7$ and $Yb_2Zr_2O_7$ with molten V_2O_5 can be expressed as follows, respectively:

$$V_2O_5(l) + SmYbZr_2O_7(s) \rightarrow SmVO_4(s) + YbVO_4(s) + 2m\text{-}ZrO_2(s) \qquad (3)$$

$$V_2O_5(l) + Yb_2Zr_2O_7(s) \rightarrow 2YbVO_4(s) + 2m\text{-}ZrO_2(s) \qquad (4)$$

4.2 Hot corrosion mechanism in pure sulfate melt

Fluxing mechanisms were proposed to interpret the hot corrosion behavior including sulphur in previous studies.[21] Similar to the reactions between vanadium compounds and solid solutions of metal oxides, in the chemical interaction of sulfate melt with rare earth zirconates, the acid-base properties of solid solutions of metal oxides were decided by the relative acid-base strength of the involved reactants.[21] As revealed in Fig. 8, no hot corrosion interactions occur between Na_2SO_4 and $(Sm_{1-x}Yb_x)_2Zr_2O_7$, and therefore here take $Sm_2Zr_2O_7$ as an example to illustrate the hot corrosion mechanism. Since $Sm_2Zr_2O_7$ can be categorized as either strong acidic oxide former or the strong basic oxide depending on the molten salts, the hot corrosion behavior of $Sm_2Zr_2O_7$ could be explained based on the fluxing mechanism. At the testing temperature, sodium sulfate decomposes according to the following equation:[21]

$$Na_2SO_4 = Na_2O + SO_3 \quad \log K = -16.7 \text{ (at 900 °C)} \tag{5}$$

This melt exhibits acid–base property, with the acidic component SO_3 and the basic component Na_2O. As shown in Eq. (5), the activity of Na_2O (a_{Na_2O}) and the pressure of SO_3 (P_{SO_3}) have the opposite tendency. During the hot corrosion test, the activity of Na_2O or the pressure of SO_3 in the molten salt determines the mechanism and extent of reaction.[12] Fluxing of oxides may work as either basic or acidic solutes in the molten salt during hot corrosion. In case of the high Na_2O activity, $Sm_2Zr_2O_7$ can react with Na_2O and dissolve in the molten sulfate by basic fluxing. Otherwise, $Sm_2Zr_2O_7$ can be dissolved by acidic fluxing in case of the low Na_2O activity and the correspondingly high P_{SO_3}. However, if the activity of Na_2O and the pressure of SO_3 are within the intermediate range, $Sm_2Zr_2O_7$ is stable and no chemical reaction occurs. The result in this study is just consistent with the last situation, and that is the reason why corrosion products are not detected.

4.3 Hot corrosion mechanism in sulfate-vanadate melt

From the phase diagram of Na_2SO_4–V_2O_5 binary system,[28] Na_2SO_4 and V_2O_5 with a molar ratio of 1:1 will react at the testing temperature in the following approach:

$$Na_2SO_4 (l) + V_2O_5 (l) \rightarrow 2NaVO_3 (l) + SO_3 (g) \tag{6}$$

According to previous research,[29] the presence of $NaVO_3$ enhances the acidic solubility, and as a result, $Sm_2Zr_2O_7$ will suffer acidic fluxing, which can be expressed by

$$2NaVO_3 (l) + Sm_2Zr_2O_7 (s) \rightarrow 2ZrV_2O_7 (s) + 2SmVO_4 (s) + Na_2O \tag{7}$$

Besides, as can be seen in the phase diagram of ZrO_2–V_2O_5 binary system, ZrV_2O_7 melts incongruently at about 747 °C to form m-ZrO_2 and a liquid mixture of m-ZrO_2 and V_2O_5.[28] When V_2O_5 is consumed, the final reaction products are m-ZrO_2 and $SmVO_4$. The introduction of vanadium into sulfate leads to subsequent formation of $NaVO_3$, which is acidic enough to dissolve $Sm_2Zr_2O_7$ by acidic fluxing. Consequently, in this case the hot corrosion mechanism can be described as

$$2NaVO_3 (l) + Sm_2Zr_2O_7 (s) \rightarrow 2 SmVO_4 (s) + 2m\text{-}ZrO_2 (s) + Na_2O (l) \tag{8}$$

As shown in Fig. 11, the difference in surface morphology of $SmVO_4$ crystals is noticed at various temperatures. At 900 °C, $SmVO_4$ crystals are block-shaped in shape; however, they exhibit a rod-like shape at temperatures of 1000–1100 °C. The morphological change from the original block-shaped crystal to the rod-like product indicates the preferential growth of $SmVO_4$ at higher temperatures. This may result from its inherent growth habit. Its growth orientation could be along a certain axis (parallel to the preferential growth direction), which means that the crystals grow faster in this direction than others. At elevated temperatures, more nuclei were formed and then selectively absorbed onto the crystallographic planes. Therefore the crystal rapidly grows preferentially along this direction. The crystallographic principles on the formation of rod-like $SmVO_4$ structure are investigated in the following section.

4.4 Crystallographic study of rod-like $SmVO_4$ hot corrosion products

The obtained $SmVO_4$ particles show well-crystallised rod-like shapes when the corrosive agent is a $Na_2SO_4+V_2O_5$ mixture. Ostwald ripening is the dominant mechanism of the nanorods growth: the formation of tiny crystalline nuclei in a supersaturated medium occurs at first, and this is followed by crystal growth.[30] The larger particles grow at the cost of the small particles; reduction in surface energy is the primary driving force for crystal growth and morphology evolution, due to the difference in solubility between the larger particles and the small particles, according to Gibbs-Thomson law.[31] The mechanism of nucleation and growth of rod-like $SmVO_4$ is a complex process of simultaneous chemical reactions and self-assembly and the schematic illustration is shown in Fig. 12. According to Rapp, the molten $NaVO_3$ works as acidic fluxing and dissolves the TBC ceramic materials.[21] During such process, the chemical reaction Eq. (8) occurs. The concentration of m-ZrO_2 and $YbVO_4$ increases with the reaction Eq. (8) proceeding. The nucleus of m-ZrO_2 and $SmVO_4$ will form when the concentrations of m-ZrO_2 and $SmVO_4$ reach the corresponding threshold concentration of the nucleus formation. At this time, the concentrations of m-ZrO_2 and $SmVO_4$ decrease and subsequently go below the threshold concentration. The nucleus of m-ZrO_2 as well as $SmVO_4$ stop forming and the growth of m-ZrO_2 and $SmVO_4$ crystals begin. With the crystals growing, the hot corrosion products will precipitate in the molten salts. As the reaction continues, the irregular nanoparticles vanish and longer microrods form. Therefore the formation progress of $SmVO_4$ microrod can be described to three consecutive steps: (i) formation of $SmVO_4$ primary particles or subunits in the early stage, (ii) aggregation of $SmVO_4$ primary particles into microrod driven by the minimization of surface energy, and (iii) conversion to loosely stacked $SmVO_4$ micro-rod via crystal aging and recrystallization. At the early reaction stage, primary $SmVO_4$ nanoparticles are formed through conventional nucleation and a subsequent crystal growth process. During the dissolution-precipitation process, to lower the free energy of the system, the original $SmVO_4$ crystal would work as seeds and the newly obtained $SmVO_4$ from reaction Eq. (8) would nucleate on the facets with the highest surface free energy. As is reported, for the zircon structure, the free energy of the facets follows such order: $(001) < (110) < (112)$.[32] In that case, the units will accumulate on the (001) plane, which means that the crystal will grow preferentially along c-axis and exhibit a rod-like morphology.

Small equiaxed hot corroded $SmVO_4$ products are obtained from V_2O_5-coated $Sm_2Zr_2O_7$ samples at different temperatures of 900–1100 °C for 2 h. This difference in grain morphology is related to the difference in the amount of seeds. Lewis basicity reflects the thermodynamic tendency of a substance to act as Lewis base. Therefore, the basicity difference is the indicative information of corrosion resistance from the viewpoint of thermodynamic factor. According to the results reported by Jones, molten V_2O_5 is believed to be more acidic than $NaVO_3$, and therefore the driving force of reaction Eq. (2) is higher than that of reaction Eq. (8). A large amount of seeds form immediately, and a high nucleation rate suppresses the growth of $SmVO_4$ rods. Chen and Rosenflanz emphasized that the driving force for formation of elongated $SmVO_4$ rods is an important factor: the lower the driving force, the lower the nucleation rates and the higher probability of formation of elongated grains.[34] As a result, a more acidic fluxing provides higher driving force and results in a subsequently higher nucleation rate. The development of elongated $SmVO_4$ grains is hampered if abundant nuclei are available, and thus the kinetics of crystal growth is so slow. That is the reason why $SmVO_4$ obtained in these two hot corrosion reactions shows distinct features. The larger driven force in the hot corrosion of V_2O_5-coated $Sm_2Zr_2O_7$ contributes to the granular $SmVO_4$. However, the relatively smaller driven force in the hot corrosion of $(Na_2SO_4+V_2O_5)$-coated $Sm_2Zr_2O_7$ is responsible for the rod-like $SmVO_4$.

5. CONCLUSIONS

Thermal conductivity of $(Sm_{1-x}Yb_x)_2Zr_2O_7$ ceramics is located within the range of 1.40–1.99 $W \cdot m^{-1} \cdot K^{-1}$ from room temperature to 1400 °C. The thermal conductivity of $SmYbZr_2O_7$ ceramic is lower than that of $Sm_2Zr_2O_7$ or $Yb_2Zr_2O_7$ ceramic. With increasing Yb content, thermal expansion coefficients of $(Sm_{1-x}Yb_x)_2Zr_2O_7$ ceramics gradually decrease at identical temperature levels. According to hot corrosion tests between pressureless-sintered $(Sm_{1-x}Yb_x)_2Zr_2O_7$ (x = 0, 0.5, 1.0) ceramics and three corrosive agents including V_2O_5, Na_2SO_4, and a $V_2O_5+Na_2SO_4$ mixture, different reaction products of ZrV_2O_7, $LnVO_4$ and m-ZrO_2 are identified depending upon the hot corrosion conditions, for example, ZrV_2O_7 and corresponding $LnVO_4$ at 600 °C for 2 h and 8h, namely $SmVO_4$, $(Sm,Yb)VO_4$, $YbVO_4$, respectively; ZrV_2O_7, m-ZrO_2 and $LnVO_4$ at 700 °C for 2 h; m-ZrO_2 and $LnVO_4$ either at 800~1100 °C for 2 h or at 700~1100 °C for 8 h. No chemical interactions are found on Na_2SO_4-coated $(Sm_{1-x}Yb_x)_2Zr_2O_7$ at 900~1100 °C. However, m-ZrO_2 and corresponding $LnVO_4$ are observed after $(Sm_{1-x}Yb_x)_2Zr_2O_7$ exposed to $Na_2SO_4+V_2O_5$ (mole ratio = 1:1) at temperatures of 600~1100 °C. Those results could be explained reasonably through phase diagram theory. The morphologies of reaction products $LnVO_4$ were rationalized by the principles of crystal growth.

ACKNOWLEDGEMENTS

This work was financially supported by the National Natural Science Foundation of China (NSFC, Grant Nos. 50972030, 51002038 and 51021002) and the Fundamental Research Funds for the Central Universities (Grant No. HIT.BRET1.2010006).

REFERENCES

1. G. Suresh, G. Seenivasan, M. V. Krishnaiah, and P. S. Murti, "Investigation of the Thermal Conductivity of Selected Compounds of Lanthanum, Samarium and Europium," *J. Alloys Compd.*, 269 L9–L12 (1998).
2. R. Vassen, X. Q. Cao, F. Tietz, D. Basu, and D. Sther, "Zirconates as New Materials for Thermal Barrier Coatings," *J. Am. Ceram. Soc.*, 83 2023–2028 (2000).
3. J. Wu, X. Wei, N. P. Padture, P. G. Klemens, M. Gell, E. Garcia, P. Miranzo, and M. I. Osendi, "Low Thermal Conductivity Rare-Earth Zirconates for Potential Thermal Barrier Coating Applications," *J. Am. Ceram. Soc.*, 85 3031–3035 (2002).
4. H. Lehmann, D. Pitzer, G. Pracht, R. Vassen, and D. Stöver, "Thermal Conductivity and Thermal Expansion Coefficients of the Lanthanum Rare-Earth-Element Zirconate System," *J. Am. Ceram. Soc.*, 86 1338–1344 (2003).
5. Z. -G. Liu, J. -H. Ouyang, Y. Zhou, J. Li, and X. -L. Xia, "Densification, Structure, and Thermophysical Properties of Ytterbium–Gadolinium Zirconate Ceramics," *Int. J. Appl. Ceram. Technol.*, 6 485–491 (2009).
6. Z. -G. Liu, J. -H. Ouyang, Y. Zhou, Q. -C. Meng, and X. -L. Xia, "Order–Disorder Transition and Thermal Conductivity of $(Yb_xNd_{1-x})_2Zr_2O_7$ Solid Solutions," *Philos. Mag.*, 89 553–564 (2009).
7. K. L. Luthra and H. S. Spacil. "Impurity Deposits in Gas-Turbines from Fuels Containing Sodium and Vanadium," *J. Electrochem. Soc.*, 129 649–656 (1982).
8. R. L. Jones, "High Temperature Vanadate Corrosion of Yttria–Stabilized Zirconia Coatings on Mild Steel," *J. Am. Ceram. Soc.*, 37 271–284 (1989).
9. D. W. Susnitzky, W. Hertl, and C. B. Carter, "Destabilization of Zirconia Thermal Barriers in the Presence of V_2O_5," *J. Am. Ceram. Soc.*, 71 992–1004 (1988).
10. R. L. Jones and R. F. Reidy, "Vanadate Hot Corrosion Behavior of India, Yttria-Stabilized Zirconia," *J. Am. Ceram. Soc.*, 76 2660–2662 (1993).
11. B. R. Marple, J. Voyer, M. Thibodeau, D. R. Nagy, and R. Vassen, "Hot Corrosion of Lanthanum Zirconate and Partially Stabilized Zirconia Thermal Barrier Coatings," *J. Eng. Gas Turb. Power*, 128 144–152 (2006).

12. Z. H. Xu, L. M. He, R. D. Mu, S. M. He, G. H. Huang, and X. Q. Cao, "Hot Corrosion Behavior of Rare Earth Zirconates and Yttria Partially Stabilized Zirconia Thermal Barrier Coatings," *Surf. Coat. Technol.*, 204 3652–3661 (2010).
13. Z. -G. Liu, J. -H. Ouyang, Y. Zhou, and X. -L. Xia, "Hot Corrosion Behavior of V_2O_5-Coated $Gd_2Zr_2O_7$ Ceramic in Air at 700–850 °C," *J. Eur. Ceram. Soc.*, 29 2423–2427 (2009).
14. Z. -G. Liu, J. -H. Ouyang, Y. Zhou, and X. -L. Xia, "Coprecipitation Synthesis and Sintering Property of $(Yb_xSm_{1-x})_2Zr_2O_7$ Ceramic Powders," *Adv. Appl. Ceram.*, 109 12–17 (2010).
15. Z. -G. Liu, J. -H. Ouyang, Y. Zhou, J. Li, and X. -L. Xia, "Influence of Ytterbium- and Samarium-Oxides Codoping on Structure and Thermal Conductivity of Zirconate Ceramics," *J. Eur. Ceram. Soc.*, 29 647–652 (2009).
16. O. Kubaschewsji, C. B. Alcock, and P. J. Spencer, Materials Thermochemistry, sixth ed., Pergamon Press, Oxford, 1993, pp. 257-323.
17. R. A. Swalin, Thermdynamics of Solids, second ed., John Wiley & Sons, New York, 1972, pp. 53-87.
18. K. Kitayama and T. Katsura, *Bull. Chem. Soc. Jpn.*, 50 889–894 (1977).
19. M. K. Reser, ed., Phase Diagrams for Ceramists–1969 supplement, Fig. 2405. *American Ceramic Society*, Columbus, OH, 1969.
20. Y. Iijima, K. Kakimoto, T. Saîtoh, T. Kato, and T. Hirayama, "Temperature and RE Elemental Dependence for ZrO_2–RE_2O_3," *Phys. C* 378–381 960–964 (2002).
21. R. A. Rapp, "Hot Corrosion of Materials: a Fluxing Mechanism," *Corros. Sci.* 44 209–221 (2002).
22. Z. Chen, J. Mabon, J. -G. Wen, and R. Trice, "Degradation of Plasma-sprayed Yttria-stabilized Zirconia Coatings via Ingress of Vanadium Oxide," *J. Eur. Ceram. Soc.* 29 1647–1656 (2009).
23. R. L. Jones, "Oxide Acid-base Reactions in Ceramic Corrosion," *High Temp. Sci.*, 27 369–380 (1988).
24. A. Gluhoi, P. Marginean, D. Lupu, E. Indrea, and A. R. Biri, "Influence of Lanthanide Oxides on the Catalytic Activity of Nickel," *Appl. Catal. A*, 232 121–128 (2002).
25. S. Li, Z.-G. Liu, and J.-H. Ouyang, "Hot Corrosion Behaviour of $Yb_2Zr_2O_7$ Ceramic Coated with V_2O_5 at Temperatures of 600–800 °C in Air," *Corros. Sci.*, 52 3568–3572 (2010).
26. R. C. Buchanan and G. W. Wolter, "Properties of Hot-Pressed Zirconium Pyrovanadate Ceramics," *J. Electrochem. Soc.*, 130 1905–1910 (1983).
27. P. Mohan, B. Yuan, T. Patterson, V. H. Desai, and Y. H. Sohn, "Degradation of Yttria-Stabilized Zirconia Thermal Barrier Coating by Vanadium Pentoxide, Phosphorous Pentoxide, and Sodium Sulfate," *J. Am. Ceram.* 90 3601–3607 (2007).
28. G. A. Kolta, I. F. Hewaidy, and N. S. Felix, *Erzmetall*, 25 327–330 (1972).
29. R. A. Rapp, Y. S. Zhang, "Hot Corrosion of Materials–Fundermental Studies," *JOM*, 46 47–55 (1994).
30. W. Fan, X. Y. Song, S. X. Sun, and X. Zhao, "Microemulsion-Mediated Hydrothermal Synthesis and Characterization of Zircon-Type $LaVO_4$ Nanowires," *J. Solid State Chem.* 180 284–290 (2007).
31. T. Wu and Y. Z. Chen, "Analytical Studies of Gibbs-Thomson Effect on the Diffusion Controlled Spherical Phase Growth in a Subcooled Medium," *Heat Mass Transf.*, 39 665–674 (2003).
32. J. D. Donnary and D. Harker, "A New Law of Crystal Morphology Extending the Law of Bravais," *Am. Miner.*, 22 446–467 (1937).
33. Y. G. Yu, Y. Cheng, H. J. Zhang, J. Y. Wang, X. F. Cheng, and H. R. Xia, "Growth and Thermal Properties of $YbVO_4$ Single Crystal," *Mater. Lett.*, 60 1014–1018 (2006).
34. I. W. Chen and A. Rosenflanz, "A Tough SiAlON Ceramic Based on α-Si_3N_4 with a Whisker-Like Microstructure," *Nature*, 389 701–704 (1997).

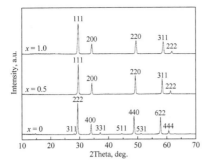

Fig. 1. XRD patterns of $(Sm_{1-x}Yb_x)_2Zr_2O_7$ ceramics sintered at 1700 °C for 10 h in air

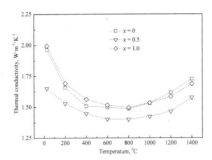

Fig. 2. Thermal conductivity of $(Sm_{1-x}Yb_x)_2Zr_2O_7$ ceramics as a function of temperature.

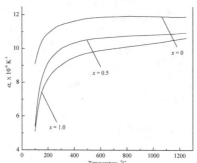

Fig. 3. Thermal expansion coefficients of $(Sm_{1-x}Yb_x)_2Zr_2O_7$ ceramics as a function of temperature.

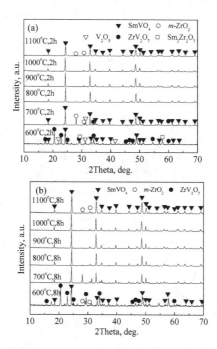

Fig. 4. XRD patterns of V_2O_5-coated $Sm_2Zr_2O_7$ samples heat-treated at temperatures of 600–1100 °C for different holding time in air: (a) 2 h, (b) 8 h.

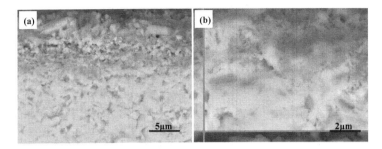

Fig. 5. Backscattered electron images of the cross-sections of V_2O_5-coated $(Sm_{1-x}Yb_x)_2Zr_2O_7$ specimens heat-treated at 800 °C for 8 h in air: (a) $SmYbZr_2O_7$, (b)$Yb_2Zr_2O_7$.

Fig. 6. Surface morphologies of hot-corroded products on V_2O_5-coated $Sm_2Zr_2O_7$ samples heat-treated at different temperatures for 2 h in air: (a) 600 °C; (b) 800 °C, (c) 700 °C.

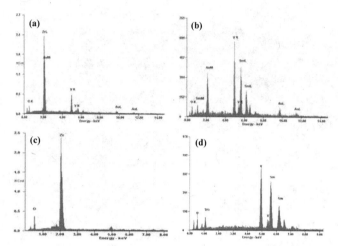

Fig. 7. EDS analysis of the positions A, B, C and D marked in Fig. 6: (a) position A; (b) position B; (c) position C; (d) position D.

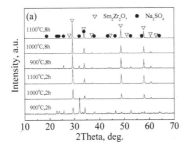

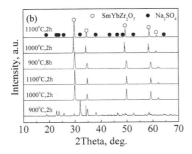

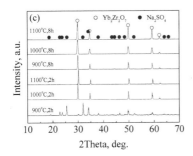

Fig. 8. XRD patterns of Na$_2$SO$_4$-coated (Sm$_{1-x}$Yb$_x$)$_2$Zr$_2$O$_7$ heat-treated at temperatures of 900–1100 °C for 8 h in air: (a) Sm$_2$Zr$_2$O$_7$, (b) Yb$_2$Zr$_2$O$_7$, (c) SmYbZr$_2$O$_7$.

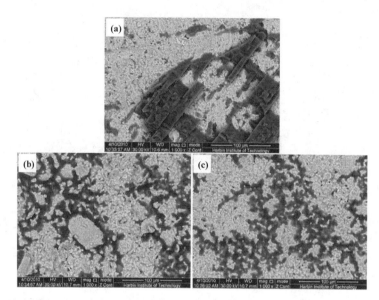

Fig. 9. Backscattered electron images of surface morphologies of Na_2SO_4-coated $(Sm_{1-x}Yb_x)_2Zr_2O_7$ specimen heat-treated at 900 °C for 8 h in air: (a) $x = 0$, (b) $x = 0.5$, (c) $x = 1.0$.

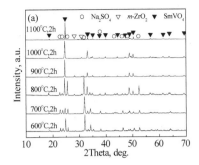

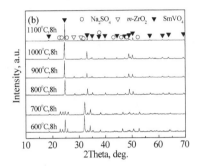

Fig. 10. XRD patterns of $Na_2SO_4+V_2O_5$-coated $Sm_2Zr_2O_7$ specimen heat-treated at 900–1100 $^{\circ}$C in air: (a) 2 h, (b) 8 h.

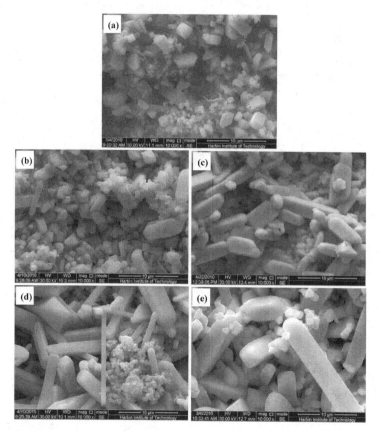

Fig. 11. Surface morphology of $(Na_2SO_4+V_2O_5)$-coated $Sm_2Zr_2O_7$ heat-treated at 700–1100 °C for 2 h in air: (a) 700 °C, (b) 800 °C, (c) 900 °C, (d) 1000 °C, (e) 1100 °C.

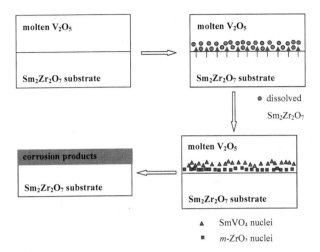

Fig. 12. A schematic plot to illustrate the hot corrosion mechanism of $Sm_2Zr_2O_7$ against V_2O_5 and $Na_2SO_4+V_2O_5$ mixture.

Table 1 A summary table reflecting the presence of the following phases in XRD results of V_2O_5-coated $Yb_2Zr_2O_7$ samples heat-treated at temperatures of 600–1100 °C for different holding time.

	YbVO$_4$	m-ZrO$_2$	ZrV$_2$O$_7$	Yb$_2$Zr$_2$O$_7$
600°C/2h	√	—	√	√
600°C/8h	√	—	√	√
700°C/2h	√	√	√	√
700°C/8h	√	√	—	√
800°C/2h	√	√	—	√
800°C/8h	√	√	—	√
900°C/2h	√	√	—	√
900°C/8h	√	√	—	√
1000°C/2h	√	√	—	√
1000°C/8h	√	√	—	√
1100°C/2h	√	√	—	√
1100°C/8h	√	√	—	√

Table 2 A summary table reflecting the presence of the following phases in XRD results of V_2O_5-coated $SmYbZr_2O_7$ samples heat-treated at temperatures of 600–1100 °C for different holding time.

	(Sm,Yb)VO$_4$	m-ZrO$_2$	ZrV$_2$O$_7$	SmYbZr$_2$O$_7$
600°C/2h	√	—	√	√
600°C/8h	√	—	√	√
700°C/2h	√	√	√	√
700°C/8h	√	√	—	√
800°C/2h	√	√	—	√
800°C/8h	√	√	—	√
900°C/2h	√	√	—	√
900°C/8h	√	√	—	√
1000°C/2h	√	√	—	√
1000°C/8h	√	√	—	√
1100°C/2h	√	√	—	√
1100°C/8h	√	√	—	√

VARIATION OF CREEP PROPERTIES AND INTERFACIAL ROUGHNESS IN THERMAL
BARRIER COATING SYSTEMS

P. Seiler, M. Bäker and J. Rösler
Institute for Materials (IfW), Technische Universität Braunschweig
Braunschweig, Germany

ABSTRACT
    A typical thermal barrier coating system consist of the bond-coat (BC) the thermal barrier coat-
ing (TBC) and a thermally grown oxide (TGO) between the bond-coat and TBC. A simplified coating
system is introduced and simulated which consists of a MCrAlY bond-coat material as the substrate, an
TGO, and a TBC on top. The influence of the nickel-based substrate can be neglected, which reduces the
influencing parameters. The failure mechanism can be analyzed by varying the creep properties of the
bond-coat material (fast creeping Fecralloy and slow creeping ODS strengthened MA956). The influence of the interfacial roughness can be examined varying the amplitude and wavelength of the interface.
It is shown that a fast creeping bond-coat benefits the lifetime of the coating system. Different FEM
simulations of the coatings support this assumption.

INTRODUCTION
    Thermal barrier coating systems are used on top of the highly stressed nickel-based turbine
blades of gas turbines. The coatings protect the underlying substrates from the 1400°C hot gas temperature. A standard coating system can be found in Figure 1(a). It consists of a nickel-based material and
two additional layers: the metallic bond-coat and the ceramic thermal barrier coating (TBC). A third
layer is formed in service between the TBC and the bond-coat: the so called thermally grown oxide
(TGO). The TGO forms because oxygen diffuses through the TBC due to the high porosity of the TBC
and the high ionic diffusivity of oxygen in zirconia [1, 2]. The coating system fails in service because of
the growth of the TGO, the interfacial roughness [3], creep processes, sintering processes in the TBC,
the complex load conditions [2], and stresses in the coating which are induced by the mismatch of the
thermal expansion coefficient of all layers [3, 4].
    A model system was developed which reduces the influencing parameters of the coating sys-
tem (Figure 1(b)). The model system consists of the MCrAlY bulk material Fecralloy which is in this
case used as the substrate. Fecralloy consists of 72.8 wt % iron, 22 wt % chromium, 5 wt % aluminium,
0.1 wt % yttrium, and 0.1 wt % zirconium. The model system allows to study the creep influence of the
bond-coat material itself by using different MCrAlY bulk materials. Therefore, a second substrate was
used in the model system: MA956 as a slow creeping oxide-dispersion-strengthened (ODS) material.
The Y_2O_3 partially stabilized plasma sprayed zirconia TBC is applied by atmospheric plasma spaying
directly to the substrate. Therefore, the failure mechanisms in the bond-coat material and the TBC can
be studied without the influence of the nickel-based materials.
    The oxidation at high temperature is crucial for the delamination of the coating system. The
failure mechanisms of the simplified model are comparable to a standard thermal barrier coating system
due to the same oxidation behavior. However, the nickel-based substrate has an influence on the coating
system due to diffusion, phase transformations in the substrate and stress component in axial direction
(11 direction in Figure 1(b)) as shown in [5]. Thus, the results obtained by the model system have to be
checked carefully with standard thermal barrier coating system.

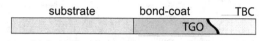

(a) Standard coating system on top of a nickel-based substrate.

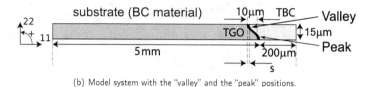

(b) Model system with the "valley" and the "peak" positions.

Figure 1: Sketches of a standard coating system and the used model system.

In this paper different FE models were developed to study the failure mechanisms, the creep influence of the bond-coat material and the influence of the bond-coat roughness which are described in the following.

FEM MODEL

The geometry is based on [3, 4, 6]. It was calculated with the ABAQUS finite element code. Figure 1(b) shows a sketch of the model. It consists of CAX4HT elements with a linear approximation. The substrate/TGO interface and the TGO/TBC interface, respectively, are approximated with a sinusoidal function. The thickness of the TGO is constant at every point normal to this interface. The FEM model is axially symmetric and contains symmetric boundary conditions. Thus, an infinite cylinder is simulated with a fully periodic sinusoidal substrate/TGO and TGO/TBC interfaces.

The analysis of the stresses is performed at prominent positions at the interface (valley and peak positions in Figure 1(b)). The discussed stress values at this positions are the mean values of the radial stresses σ_{11} in the selected elements at the integration points.

The measured parameters of the substrate materials Fecralloy and MA956 are shown in Table 1. Fecralloy and MA956 are compared in the following by varying the creep parameters. It is assumed that both materials have the same mechanical properties except the creep parameters. Thus, the influence of creep can be studied.

The material parameters of the TGO and the TBC can be found in [3]. The creep properties at 1000°C of the TGO and TBC are varied. The slow and fast creeping parameters of these layers can also be found in [3]. The creep rate in the TBC and in the TGO was calculated by a Norton creep law

$$\dot{\varepsilon} = A\sigma^n. \tag{1}$$

The creep rate of Fecralloy and MA956 is calculated by

$$\dot{\varepsilon} = \underbrace{A_0 \cdot \exp(aT) \cdot \exp\left(-\frac{Q}{RT}\right)}_{A} \cdot \sigma^n \tag{2}$$

with the activation energy Q, the creep exponent n and an additional factor a (Table 1) [7]. The creep parameters Q, A_0, n, and a are fitted values obtained from experimental creep tests. The creep behaviour of MA956 is characterized by three regimes with independent sets of creep parameters. The transition from one regime to the other takes place at the critical stress σ_{c1} and σ_{c2} (Table) [8]. They are calculated during the simulation by equating two equations (2) with the different creep parameters in the two regimes. For instance the first critical stress is defined as

$$\sigma_{c1} = \left(\frac{A_1}{A_2} \right)^{\frac{1}{n_2 - n_1}}.$$ (3)

$A_1(A_0, Q, a, R, T)$ and n_1 are the parameters in the range $\sigma < \sigma_{c1}$; $A_2(A_0, Q, a, R, T)$ and n_2 are the parameter in the range $\sigma_{c1} < \sigma < \sigma_{c2}$, respectively.

Table 1: Material parameters of Fecralloy and MA956.
(*)The creep parameters of MA956 depend on the current stress (see Table).

	Fecralloy	MA956
Young's modulus E, 20°C [GPa]	177	177
Young's modulus E, 1000°C [GPa]	95	95
ν, 20°C	0.30	0.30
ν, 1000°C	0.33	0.33
α, 20°C [K^{-1}]	1.11×10^{-5}	1.11×10^{-5}
α, 1000°C [K^{-1}]	1.46×10^{-5}	1.46×10^{-5}
density ρ [kg/m^3]	7.1×10^3	7.1×10^3
creep activation energy Q [kJ/mol]	486	depends on σ(*)
creep exponent n [-]	5.29	depends on σ(*)
creep prefactor A_0 [MPa^{-n}s^{-1}]	1.72×10^{-10}	depends on σ(*)
creep factor a [K^{-1}]	-0.012	depends on σ(*)

Table 2: Creep parameters of MA956 which are used in Eq. (2).

	A_0 [MPa^{-n}s^{-1}]	n [-]	Q [kJ/mol]	a [K^{-1}]
$\sigma < \sigma_{c1}$	78.978	4.9827	453	0.0
$\sigma_{c1} < \sigma < \sigma_{c2}$	3.466×10^{-124}	41.0	453	0.1
$\sigma > \sigma_{c2}$	8.68×10^{16}	5.2911	486	-0.0122

The TGO growth was simulated by anisotropic swelling of the whole TGO with a 10 times larger growth rate normal to the interface compared to the direction parallel to the interface [4]. A Tammann law was used to calculate the anisotropic growth rate of the TGO [9] by

$$\dot{s} = \frac{1}{2} \cdot \frac{k'_p}{s}$$ (4)

with the current TGO thickness s, the parabolic oxidation constant k'_p and the growth rate $\dot{s}$. The parabolic oxidation constant of this simulation is $k'_p = 1.5 \times 10^{-17}$ m^2/s which describes the oxidation kinetics of α-Al$_2$O$_3$ at 1000°C [9]. The growth rate is calculated during every simulated thermal cycle. The growth rate depends on the current TGO thickness which is measured at the edge of the simulated

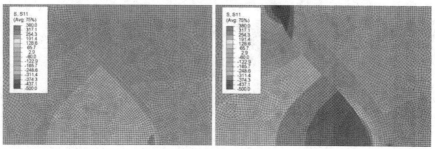

(a) 1st thermal cycle, TGO thickness: 0.50 μm   (b) 50th thermal cycle, TGO thickness: 2.54 μm

Figure 2: Anisotropic oxide swelling in the Fecralloy system (σ_{11} in MPa, see Figure 1(b)). The figures show the stress states after the first and after 50 full thermal cycle with a fast creeping TBC and TGO.

value near the peak position (Figure 1(b)). It is assumed that the model is in a stress free state at RT at the beginning of the simulation.

A preprocessor for ABAQUS was written which ensures the meshing of different regions in the model. The model is re-meshed at the beginning of every thermal cycle by defining the element size in every region of the model in advance. For instance, the minimum element size inside the TGO and near the interfaces is $\sim 0.25\,\mu m$ at the beginning of every thermal cycle. The interpolation from the old to the remeshed model is calculated by ABAQUS [10].

CREEP

The radial stress fields in the model at the beginning of the simulation and after 50 thermal cycles are shown in Figure 2. Every thermal cycle includes the heating process, dwelling for 2 h at 1000°C and cooling to RT. The radial stresses σ_{11} at the substrate/TGO and the TGO/TBC interfaces against the total simulation time with a fast and slow creeping TBC and TGO is shown in Figure 3 and Figure 4.

It can be found that the stresses in the MA956 substrate are always larger than the stresses in the Fecralloy substrate. This is due to the higher creep rate in the Fecralloy substrate. The compression and tension areas inside the TBC starts to shift after 4 cycles if a fast creeping TGO and a fast creeping TBC is simulated (Figure 3). The stress shift results from the growing influence of the stresses induced by the TGO due to the different thermal expansion coefficients.

On the other hand, the TBC stresses in the Fecralloy system are larger than the TBC stresses in the MA956 coating system during all thermal cycles (Figure 3(b)). This results in a changed geometry of the TGO. The growth of the TGO in lateral direction results in a changed position of the TGO in which the valley of the TGO is shifted to the inside (in Figure 2 to the left) and the peak to the outside (in Figure 2 to the right). The comparatively fast creeping Fecralloy substrate can not resist the strain because of the resulting low stresses. The resistance of the MA956 substrate to the strain of the TGO is larger caused by the creep strength. This leads to higher magnitude of stresses inside the substrate and to lower stresses inside the TBC in the MA956 system.

The slow creeping TBC and the slow creeping TGO show the same behavior (Figure 4). The magnitude of stresses differ because not all stresses can relax during the dwelling at high temperature. Therefore, the growth stresses of the TGO influence the coating.

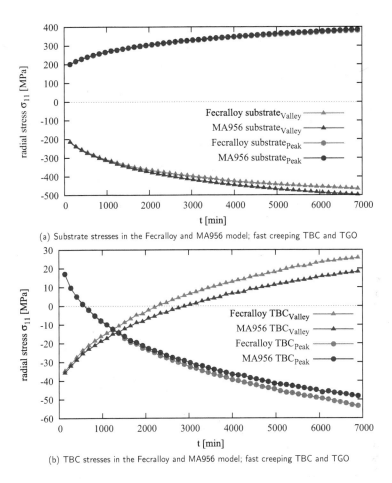

(a) Substrate stresses in the Fecralloy and MA956 model; fast creeping TBC and TGO

(b) TBC stresses in the Fecralloy and MA956 model; fast creeping TBC and TGO

Figure 3: Radial stresses at the end of every thermal cycle in the valley and peak position of the model with a fast creeping TBC and TGO.

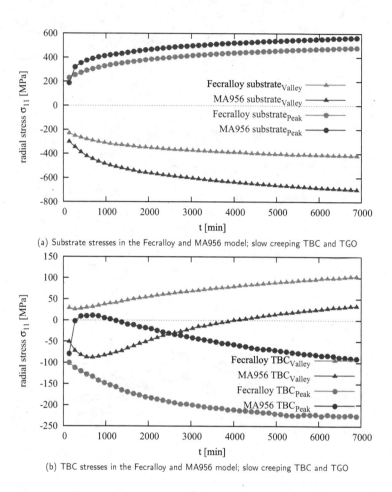

(a) Substrate stresses in the Fecralloy and MA956 model; slow creeping TBC and TGO

(b) TBC stresses in the Fecralloy and MA956 model; slow creeping TBC and TGO

Figure 4: Radial stresses at the end of every thermal cycle in the valley and peak position of the model with a slow creeping TBC and TGO.

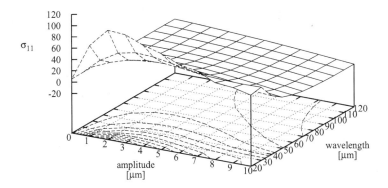

Figure 5: Elastic cooling stresses in the TBC valley 2.0 μm TGO thickness, σ_{11} in MPa) of different sinusoidal interfaces.

The simulation results show that the Fecralloy system fails more likely above the TGO (at the TGO/TBC interface) and the MA956 system under the TGO (at the substrate/TGO interface) due to higher stresses in this areas.

INTERFACIAL ROUGHNESS

The interfacial roughness was studied by varying the wavelength and the amplitude of the substrate/TGO interface. Only the cooling stresses were taken into account in this simulations. The model is in a stress free state at high temperature. Furthermore, the TGO thickness is constant for the whole simulation.

Figure 5 shows the results of the elastic simulation in the valley of the TBC without varying the creep parameters. Cooling stresses up to 100 MPa can be found at low wavelengths of the interface. In this range the TGO is thick compared to the roughness. Therefore, the TGO has a profound influence on the stress state compared to higher wavelengths. The same behaviour can be found at low amplitudes: the TGO influences the stress state at low amplitudes until the amplitude reaches 3 μm. At higher amplitudes the bond-coat dominates the stress state in the TBC.

The simulations are pure elastic without creep or plastic deformation. The stresses depend only on the wavelength, the amplitude and the TGO thickness. Hence, the TGO dominates the stress state, if the oxide layer is thick compared to the wavelength and the amplitude of the interfacial roughness. Figure 5 suggests a lifetime optimum at high wavelength and amplitudes if the TGO thickness is thin compared to the amplitude and wavelength.

CONCLUSION

A real TBC coating system consists of the TBC, TGO and bond-coat. A slow creeping bond-coat may reduce the lifetime of the system because of cracks under the TGO at the bond-coat/TGO interface. On the other hand, the slow creeping bond-coat reduces the stresses inside the TBC. Thus, optimized bond-coat creep properties may lead to a lifetime optimum of the whole coating system.

The interfacial roughness of the bond-coat plays also a crucial role. It was shown that the stresses near the interface depend on the wavelength and the amplitude of the interfacial roughness. Furthermore, the TGO influences the stresses state if the TGO is thick compared to the wavelength and the amplitude of the interfacial roughness. In this case the TGO dominates the stresses near the interface. A lifetime optimum of the coating systems may be found at high wavelengths and amplitudes of the bond-coat if the TGO is thin compared to the amplitude and wavelength. On the other hand, a certain bond-coat roughness is necessary to increase the mechanical bonding of the TBC. Further simulations with creep in all coating layers will show the interaction between creep, TGO thickness and the bond-coat interface.

REFERENCES

[1] Fox A C and Clyne T W *Surface and Coatings Technology* **184** 311–321 (2004)

[2] Padture N P, Gell M and Jordan E H *Science* **296** 280 – 284 (2002)

[3] Bäker M, Rösler J and Heinze G *Acta Materialia* **53** 469–476 (2005)

[4] Rösler J, Bäker M and Aufzug K *Acta Materialia* **52** 4809–4817 (2004)

[5] Bäker M, Rösler J and Affeldt E *Computational Materials Science* **47** 466–470 ISSN 0927-0256 (2009)

[6] Seiler P, Bäker M, Beck T, Schweda M and Rösler J *Journal of Physics: Conference Series (Proceedings ICSMA-15)* (2009)

[7] Herzog R, Bednarz P, Trunova E, Shemet V, Steinbrech R W, Schubert F and Singheiser L *Ceramic Engineering and Science Proceedings* **27** (2006)

[8] Herzog R, Schuster H, Schubert F and Nickel H (1994) *Mikrostruktur und mechanische Eigenschaften der Eisenbasis-ODS-Legierung PM2000* Ph.D. thesis Insitut für Werkstoffe und Energietechnik, FZ Jülich

[9] Bürgel R (2001) *Handbuch Hochtemperatur- Werkstofftechnik* 2nd ed (Braunschweig/ Wiesbaden: Vieweg)

[10] Bäker M *Computational Materials Science* **43** 179–183 ISSN 0927-0256 proceedings of the 16th International Workshop on Computational Mechanics of Materials - IWCMM-16 (2008)

Materials for Extreme Environments: Ultra High Temperature Ceramics (UHTCS) and Nanolaminated Ternary Carbides and Nitrides (Max Phases)

TEMPERATURE AND STRAIN-RATE DEPENDENT PLASTICITY OF ZrB$_2$ COMPOSITES FROM HARDNESS MEASUREMENTS

J. Wang*, F. Giuliani*, L.J. Vandeperre*
* UK Centre for Advanced Structural Ceramics and Department of Materials, Imperial College London, South Kensington Campus, London SW7 2AZ, UK

ABSTRACT

In this work, nanoindentation was used to determine the strain-rate and temperature dependence of the hardness of a ZrB$_2$ composite to derive constitutive relations for plasticity controlled by the lattice resistance. The lattice resistance or Peierls' stress was estimated as 6.9±0.3 GPa with an average activation volume of 2.3±0.8 × 10^{-29} m^3. Transmission electron microscopy of the deformation underneath the indents is consistent with slip on both basal as well as prismatic planes.

INTRODUCTION

Zirconium diboride (ZrB$_2$) and ceramic composites based on it have an excellent combination of physical and chemical properties, such as a high melting point, high strength, high hardness, high thermal and electrical conductivity and chemical stability. These properties make these materials candidates for a variety of ultra-high temperature applications[1] such as the leading edges of next generation hypersonic spacecraft and thermal protection systems[1]. Within this family of materials, ZrB$_2$ reinforced with 20 vol% silicon carbide (SiC) is commonly considered as a leading candidate giving the best combination of oxidation resistance and mechanical properties.

The mechanical properties of ZrB$_2$-based materials have been studied mostly at room temperature and there is only limited information about their mechanical and thermal properties at the intended service temperatures (> 2273 K). Our previous study has shown the hardness, which is an indication of flow stress, drops from >20 GPa at room temperature to around 13 GPa at temperatures as low as 673 K[2] in agreement with data from Atkins[3] and data from Bsenko[4]. Such a sharp decrease in hardness is quite common in carbides and borides as documented by Atkins and Tabor[3] and illustrated in Figure 1.

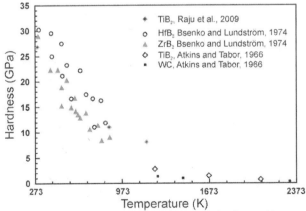

Figure 1 Selected literature data illustrating how a rapid decrease in hardness with temperature is common for borides[3-5] and carbides[3].

The rapid decrease in hardness is the result of the relative small activation energy for lattice controlled dislocation flow so that thermal activation has strong influence as discussed extensively in [6]. Hence, dislocation flow could be important at more elevated temperatures. This is supported by the observations in Hu's work that ZrB$_2$-based composites behaved plastically rather than brittle during three point bending tests at 2073 K [7] and Hagerty and Lee's finding that ZrB$_2$ showed 10% plastic strain during compression testing at 2400 K [8].

Indentation techniques are primarily used to measure hardness of materials, but can be used to study other mechanical properties such as toughness and yield strength[9, 10]. With the advent of depth-sensing indentation equipment, which record the load and depth during indentation, it is now also possible to investigate the time dependence of the hardness[11-13] and therefore the strain rate sensitivity. Given that for ceramic materials, the resistance to plastic flow at low temperature can be expected to be dominated by the lattice resistance or Peierls' stress, the aim of this paper is to obtain reasonable estimates for the constitutive parameters for lattice resistance controlled dislocation flow from measurements of the strain rate sensitivity and temperature dependence of the hardness of ZrB$_2$.

EXPERIMENTAL

Commercially available ZrB$_2$ (H.C. Stark, Grade B), SiC (H.C. Stark, Grade UF-10) and B$_4$C (H.C. Stark Grade HS) were used to prepare the material. A phenolic resin (CR-96 Novolak, Crios Resinas, Brazil) was added as a source of carbon (1 wt% carbon). Silicon carbide, carbon and boron carbide were added as they aid reaching full density for pressureless sintering[14]. The assumption was that a limited amount of secondary phases would influence the derived properties less than the porosity, which would remain in the absence of the additives.

The mixture was ball milled in methyl ethyl ketone (MEK) for 24 hours with tungsten carbide balls as milling media. The mixture was then dried using a rotary evaporator (Rotavapor R-210/215, Büchi, Germany) and pressed to 13 mm diameter pellets. The pellets were heated to 673 K for 2 hours to convert the resin into carbon and sintered without pressure in flowing Argon at 2273 K for 2 hours, reaching >99% relative density. More details on the production of these materials can be found elsewhere[2]. A typical micrograph of the microstructure is shown in Figure 2. The elongated SiC grains are the result of using a fine SiC powder together with a coarse ZrB$_2$ powder as discussed in [14].

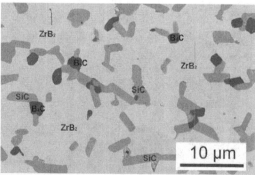

Figure 2 Scanning electron microscope micrograph (backscattered) of the ZrB$_2$ composite containing 20 vol% SiC and 2 wt% B$_4$C.

Indentation experiments were carried out at 4 temperatures between 273 K and 573 K using a Nanotest Platform 2 (Micromaterials Ltd., Wrexham, UK) with a heated Berkovich diamond and a heated sample stage, which enables measurements to be made with minimal thermal drift. The indentation schedule consisted of the following sequence: the thermal stability of the capacitive displacement sensor was measured for 1 minute before the experiment, followed by loading at a fixed rate of load increase to a predetermined maximum load. The maximum load was maintained for 60 s before unloading at the same rate as during loading. A second stability measurement was carried out during unloading at 10% of the maximum load. The measured drift of the sensor was typically 0.02 nm s^{-1} at room temperature experiment, 0.1 nm s^{-1} for 100 °C and 0.2 nm s^{-1} at 200 °C. For each temperature 3 different loading rates were used so that the maximum load was reached respectively in 4, 20 or 100 s. For data collected during loading, the apparent strain rate can be calculated from[15]

$$\varepsilon^\bullet = \frac{1}{2F}\frac{dF}{dt}$$

(1)

where F is the applied load, t is time and $\varepsilon^\bullet$ is the strain rate. Because strain rate effects relax rapidly during indentation experiments, the hardness was determined from the data collected during loading using a variant of the Oliver and Pharr method[16] in which the contact depth is determined assuming the reduced modulus remains constant during the experiment as described in full in ref.[15]. These values were then used to calculate the average hardness and average strain rate during loading. To widen the range of strain rates covered by the experiments, the data collected during the dwell at maximum load was analyzed with the same technique and related to apparent strain rate through:

$$\varepsilon^\bullet = \frac{d^\bullet}{d}$$

(2)

where d is the displacement of the indenter, $d^\bullet$ is the rate of displacement with respect to time of the indenter. Again the average strain rate and average hardness for all data collected during dwelling was determined from all data.

Each of these experiments was repeated to a maximum load of 50, 100, 200 or 500 mN. For each of the experimental variations 12 indents were made. Scanning electron microscopy (SEM, LEO Gemini 1525 FEGSEM, Carl Zeiss, Germany, accelerating voltage 20 kV) and transmission electron microscopy (TEM, 2000FX, JEOL, Japan, accelerating voltage: 200kV) were used to study the indents. The analysed data were reviewed and matched to SEM images to reject poor data (e.g. an indent made on the interface of two phases) and indents made in SiC or B$_4$C particles. To investigate the deformation around the indents, cross-sectional TEM samples through the indentations were prepared with the aid of a focus ion beam (FIB, Helios NanoLab 50 series DualBeam, FEI Company, The Netherlands) following the method developed by Langford and Pettford-Long[17],

RESULTS AND DISCUSSION

Nanoindentation Measurements
The results of the hardness measurements are plotted in Figure 3 as a function of the applied load. In agreement with earlier findings [2], the measured hardness depends on load: the hardness is 23 GPa up to 50 mN and decreases gradually towards 20 GPa for higher loads. A similar variation was obtained at higher temperatures up to 573 K. The higher hardness at the lower load of 50 mN is attributed to the absence of cracking whereas at higher loads cracks form. A similar trend for hardness values is found for all temperatures. Therefore, to minimise the effect of cracking on the estimates for the resistance to plastic flow, the data obtained from 50 mN indents were used.

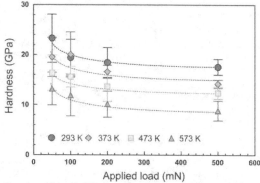

Figure 3: Hardness as a function of applied load at the different test temperatures

The results of the hardness versus strain rates at different temperatures are shown in Figure 4. The data points show the average for each of the strain rate ranges (3 on the right defined by the loading rate, 1 on the left defined by the self-equilibration during a dwell at maximum load). The hardness values as a function of strain rates can then be converted into flow stress values as a function of strain rate using the generalised relationship between hardness, H, and uni-axial yield strength, Y, derived by Vandeperre et al.[18] by modifying the expanding cavity solution of Hill[19]:

$$\frac{H}{Y} = \frac{2}{3}\left\{1 + \frac{3}{3-6\lambda}\ln\left(\frac{(3+2\mu)\cdot(2\lambda\cdot(1-\zeta)-1)}{(2\lambda\mu-3\mu-6\lambda)\cdot\zeta}\right)\right\} \tag{3}$$

with:

$$\zeta = \frac{E_r}{E_r - 2H\tan\alpha}$$

$$\lambda = \frac{(1-2\nu)Y}{E}$$

$$\mu = \frac{(1-\nu)Y}{E}$$

where E is the Young modulus, E_r is reduced modulus, ν is Poisson ratio (0.15), and α the included equivalent semi-angle of the indenter. The value of the Poisson ratio was calculated from the single crystal elastic constants given in [20]. The elastic modulus used was 499 GPa as determined from the indentation data. This value is in close agreement with the average elastic modulus calculated from the single crystal data (500 GPa).

In deriving these relationships the Tresca yield criterion is used and therefore the uni-axial yield stress obtained is twice the shear flow stress The average data was then used to construct a plot of the variation of flow stress, τ, with strain rate at the different temperatures, see Figure 5. From Figure 5 it is clear that the data for the highest strain rate obtained from indentation at 293 K and 373 K is about 500 MPa higher than expected from the trend established from indentation at higher temperatures. These two data points were therefore not used in determining the average parameters.

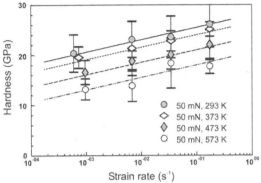

Figure 4 Hardness versus strain rates at different temperatures.

To calculate constitutive data for lattice resistance dominated plastic flow from the experimental results, a simple model for the relation between the strain rate and shear flow stress can be developed starting from the well established relation between the strain rate, the mobile dislocation density, ρ_m, and dislocation velocity, v:

$$\frac{d\gamma}{dt} = \rho_m \cdot b \cdot v \tag{4}$$

where the velocity of the dislocations can be estimated using a standard approach for a stress activated process:

$$v = \nu \cdot b \cdot \left[\exp\left(-\frac{(\tau_p - \tau)V}{kT} \right) - \exp\left(-\frac{(\tau_p + \tau)V}{kT} \right) \right] \tag{5}$$

where ν is the attempt frequency and τ_p is the lattice resistance or Peierls' stress, which is the stress needed to make the dislocations move in the absence of any thermal energy. τ is the applied shear stress, V is the activation volume, T is the temperature and k is the Boltzmann constant. Combining these equations and solving for the applied shear stress, yields the following relationship:

$$\tau = \frac{kT}{V} \sinh^{-1} \left[\frac{\overset{\bullet}{\gamma}}{2 \cdot \rho_m \cdot \nu \cdot b^2} \exp\left(\frac{\tau_p V}{kT} \right) \right] \tag{6}$$

As long as τ V remains large relative to k T, the second term in the square brackets of eq.(5) is much less than the first, and therefore assuming the activation volume can be treated as constant, the expression can be simplified to:

$$\tau = \tau_p + \frac{kT}{V} \ln\left[\frac{d\gamma}{dt} \right] - \frac{kT}{V} \ln\left[\rho_m \cdot b^2 \cdot \upsilon \right] \tag{7}$$

Therefore, a plot of the flow stress against the natural logarithm of strain rate should give a linear relationship with slope k T / V. As shown in Figure 6, the result suggests that there is a small increase in the activation volume with temperature.

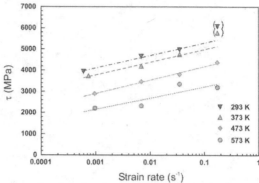

Figure 5: Shear flow stress as a function of strain rate and linear regression lines used to determine the activation volume.

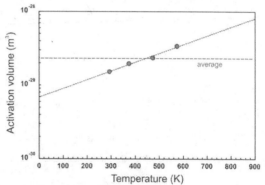

Figure 6 Variation of the activation volume with temperature.

However, in order to avoid needlessly complicating the analysis, in the first instance, the average activation volume of $2.3 \pm 0.8 \times 10^{-29}$ m^3 was used. Moreover, this average value appears reasonable: the activation volume can be expressed as $0.7 \times b^3$, with b the Burgers' vector. The magnitude of the activation volume is similar to the values reported for TiC $(0.96 \times b^3)$[6], ZrC $(1.25 \times b^3)$[6], Si $(2.86 \times b^3)$[6], Ge $(3.33 \times b^3)$[6] and MgO $(2.29 \times b^3)$[6].

An estimate for the Peierls' stress and the dislocation density was obtained by plotting the variation of the flow stress at fixed strain rate with temperature as in Figure 7. The intercept with the y-axis of the line of best fit through the data yields an estimate for the Peierls' stress, whereas the slope of such a line is predicted to be:

$$\frac{k}{V} \ln\left(\frac{\frac{d\gamma}{dt}}{\rho_m \cdot b^2 \cdot \upsilon} \right)$$

and can therefore be used to determine the mobile dislocation density using the activation volume determined before. Following Frost and Ashby[6], the attempt frequency was taken to be of the order of 10^{11} s^{-1}. With this in mind, the average estimate for the Peierls' stress is 6.9 ± 0.3 GPa, and the mobile dislocation density is estimated at $3 \pm 4 \times 10^{11}$ m^{-2}. The estimate for the dislocation density appears somewhat low relative to what would be expected for a heavily deformed material ($\sim 10^{13}$-10^{14} m^{-2})[21]. However, the dislocation density varies widely with distance from the indenter, and only an average value from the rate determining step is derived here.

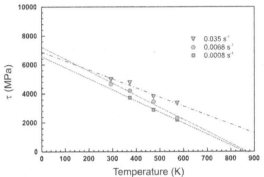

Figure 7 Variation of the flow stress at constant strain rate with temperature and the lines of best fit to the data.

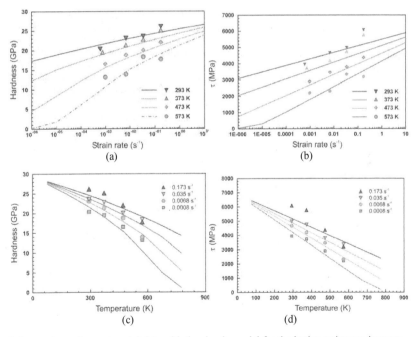

Figure 8 Comparison of experimental data with the simple model for the lattice resistance in terms of (a) hardness versus strain rate, (b) shear flow stress versus strain rate, (c) hardness versus temperature and (d) shear flow stress versus temperature.

Figure 8 compares the experimental data with the predictions obtained from the derived parameters and illustrates that a simple model using a single activation volume and a single mobile dislocation density can capture the overall variation of the measured hardness or flow stress with temperature and strain rate quite well. The assumption that τ V remains large relative to k T can now be checked: even at 573 K, τ V is two orders of magnitude larger than k T and hence the simplification was allowable.

TEM study of the active slip systems during indentation of ZrB$_2$

A bright field TEM image of the cross-section of a 50 mN indent is shown in Figure 9. The indent is in a ZrB$_2$ grain activating dislocations mainly in two planes which form an angle of 90°. As indicated in the insert, the traces are consistent with slip activity on the {0001} or basal and the {01$\bar{1}$0} or prismatic planes.

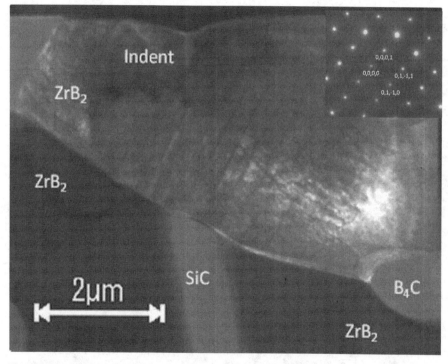

Figure 9 Transmission electron image of a cross-section of an indent made in a ZrB$_2$ grain. The selected area diffraction pattern is from the ZrB$_2$ grain in which the indentation was made.

Slip on the $\{10\bar{1}0\}$ prismatic planes is consistent with the observation of slip traces near room temperature macroscopic Knoop indents made by Haggerty and Lee[8] and near Vickers indents and scratch grooves by Ghosh et al.[22]. Slip on basal planes was also observed by Haggerty and Lee[8] but mainly during high temperature compression tests. Based on the analysis of the hardness anisotropy of Knoop indentation in ZrB₂, Nakano et al. [23] suggested that slip on the prismatic planes was dominant. However, they also concluded that substantial contributions from slip on the basal plane were needed to explain the limited hardness anisotropy observed experimentally: 24 GPa when indenting the $\{1\bar{2}10\}$ plane versus 20.6 GPa when indenting the basal plane. The limited anisotropy also suggests that the difference in the resistance to dislocation motion on these two types of planes is not very different. Hence, the estimates obtained above are probably averaged values for both slip systems.

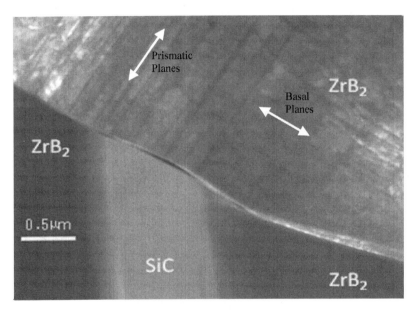

Figure 10 Higher magnification TEM image showing how the SiC grain, also shown in Figure 9, resisted the flow while the surrounding ZrB₂ material flowed away from the indentation.

The cross section also shows that the indent depth is smaller than 10% of the size of the ZrB₂ grain in the direction of indentation. In analogy with the rule of thumb for the hardness of thin films, where indents up to 10-15% of the thickness of the thin film normally yield the hardness of the film without influence of the substrate[24], it is reasonable to expect that the hardness would be the hardness of pure ZrB₂. However, Figure 9 and 10 show that the elongated SiC grain underneath the indentation appears to punch into the ZrB₂ grain. The deformation pattern is believed to have resulted from a higher resistance to plastic flow of the SiC grain relative to the surrounding ZrB₂ so that the SiC grain has not deformed, while the ZrB₂ has been displaced as material was being forced to flow away from the indenter. This means that the parameters derived here could to some extent be influenced by the presence of the SiC and B₄C. It is unfortunately difficult to quantify this as this would require that a section be taken underneath every indent. The assumption that a minor amount of other phases would not be problematic is therefore probably wrong and better estimates for the flow stress of ZrB_2 will require measurements in single crystals or in large-grained single phase materials.

CONCLUSIONS

Nanoindentation was carried out on a ZrB_2 – 20 vol% SiC composite to study strain-rate and temperature sensitivity of its hardness. The Peierls' stress and activation volume for lattice controlled dislocation flow were calculated to be 6.9 ± 0.3 GPa, and $2.3\pm0.8 \times 10^{-29}$ m^3 respectively. TEM observations suggest that dislocation flow is activated on both the basal and the prism planes. The cross-sectional TEM also indicates that when studying composites, the presence of phases underneath the indentation can influence the flow and therefore the hardness even when indents are smaller than 10% of the grain, in which they are placed. It is therefore expected that better information for the dislocation flow in ZrB_2 could be obtained from large grained, single phase materials.

ACKNOWLEDGEMENTS

Jianye Wang would like to thank the UK Engineering and Physical Sciences Research Council (EPSRC) for a PhD scholarship through the funding for the UK Centre for Advanced Structural Ceramics.

REFERENCES

1. F. Monteverde and R. Savino, Stability of ultra-high-temperature ZrB_2-SiC ceramics under simulated atmospheric re-entry conditions. *Journal of the European Ceramic Society*, **27**(16): p. 4797-4805 (2007).
2. J. Wang, F. Giuliani, and L.J. Vandeperre, The effect of load and temperature on hardness of ZrB2 composites. *Ceram. Eng. Sci. Proc.*, **31**(2): p. 59-68 (2010).
3. A.G. Atkins and D. Tabor, Hardness and deformation properties of solids at very high temperatures. *Proc. R. Soc.*, **A292**: p. 441-459 (1966).
4. L. Bsenko and T. Lundström, The high-temperature hardness of ZrB_2 and HfB_2. *Journal of the Less Common Metals*, **34**(2): p. 273-278 (1974).
5. G.B. Raju, B. Basu, N.H. Tak, and S.J. Cho, Temperature dependent hardness and strength properties of TiB_2 wtih $TiSi_2$ sinter-aid. *Journal of The European Ceramic Society*, **29**: p. 2119-2128 (2009).
6. H.J. Frost and M.F. Ashby, Deformation mechanism maps: the plasticity and creep of metals and ceramics, Oxford: Pergamon Press (1982).
7. P. Hu and Z. Wang, Flexural strength and fracture behaviour of ZrB_2-SiC ultra-high temperature ceramic composites at 1800 °C. *Journal of The European Ceramic Society*, **30**(4): p. 1021-1026 (2010).
8. J.S. Haggerty and D.W. Lee, Plastic Deformation of ZrB_2 Single Crystals. *Journal of the American Ceramic Society*, **54**(11): p. 572-576 (1971).
9. A.G. Atkins and D. Tabor, Plastic Indentation in Metals with Cones. *J. Mech. Phys. Solids*, **13**: p. 149-164 (1965).
10. Y.T. Cheng and Z. Li, Hardness obtained from conical indenters with various cone angles. *J. Mat. Res.*, **15**(12): p. 2830-2835 (2000).
11. P. Grau, G. Berg, H. Meinhard, and S. Mosch, Strain rate dependence of the hardness of glass and Meyer's law. *Journal of the American Ceramic Society*, **81**(6): p. 1557-1564 (1998).
12. N.M. Keulen, Indentation creep of hydrated Soda-Lime Silicate Glass determined by Nanoindentation. *Journal of the American Ceramic Society*, **76**(4): p. 904-912 (1993).
13. Y.I. Golovin, Y.L. Iunin, and A.I. Tyurin, Strain-rate Sensitivity of the Hardness of Crystalline Materials under Dynamic Nanoindentation. *Doklady Physics*, **48**(9): p. 505-508 (2003).

14. S.C. Zhang, G.E. Hilmas, and W.G. Fahrenholtz, Pressureless sintering of ZrB$_2$-SiC ceramics. *Journal of the American Ceramic Society*, **91**(1): p. 26-32 (2008).
15. L.J. Vandeperre, N. Ur-rehman, and P. Brown, Strain rate dependence of hardness of AlN doped SiC. *Advances in Applied Ceramics*, **109**(8): p. 493-497 (2010).
16. W.C. Oliver and G.M. Pharr, An Improved Technique for Determining Hardness and Elastic Modulus using Load and Displacement Sensing Indentation Experiments. *J. Mater. Res.*, **7**(6): p. 1564-1583 (1992).
17. R.M. Langford and A.K. Pettford-Long, Preparation of transmission electron microscopy cross-section specimens using focussed ion beam milling. *J. Vac. Sci. Techn. A*, **19**(5): p. 2186-2193 (2001).
18. L.J. Vandeperre, F. Giuliani, and W.J. Clegg, Effect of elastic surface deformation on the relation between hardness and yield strength. *Journal of Materials Research*, **19**(12): p. 3704-3714 (2004).
19. R. Hill, The Mathematical Theory of Plasticity, Oxford: Clarendon Press (1950).
20. V. Milman, B. Winkler, and M.I.J. Probert, Stiffness and thermal expansion of ZrB$_2$. *J. Phys. : Condensed Matter*, **17**: p. 2233-2241 (2005).
21. F.J. Humphreys and M. Hatherly, Recrystallization and related annealing phenomena, London: Pergamon (1996).
22. D. Ghosh, S. Ghatu, and G.R. Bourne, Room-temperature dislocation activity during mechanical deformation of polycrystalline ultra-high-temperature ceramics. *Scripta Materialia*, **61**: p. 1075-1078 (2009).
23. K. Nakano, T. Imura, and S. Takeuchi, Hardness Anisotropy of Single Crystals of IVa-Diborides. *Japanese Journal of Applied Physics*, **12**(2): p. 186-189 (1973).
24. K.W. Lee, Y.W. Chung, C.Y. Chan, I. Bello, S.T. Lee, A. Karimi, J. Patscheider, M.P. Delplancke-Ogletree, D.H. Yang, B. Boyce, and T. Buchheit, An international round-robin experiment to evaluate the consistency of nanoindentation hardness measurements of thin films. *Surface & Coatings Technology*, **168**(1): p. 57-61 (2003).

NANO-CRYSTALLINE ULTRA HIGH TEMPERATURE HfB$_2$ AND HfC POWDERS AND COATINGS USING A SOL-GEL APPROACH

S.Venugopal[1], A.Paul[1], B. Vaidhyanathan[*1], J. Binner[1], A. Heaton[2] and P. Brown[2]
[1]Department of Materials, Loughborough University, Loughborough, UK
[2] Defence Science and Technology Laboratories (DSTL), Porton Down, UK

ABSTRACT:

Nano-crystalline HfB$_2$ and HfC powders have been synthesized through a simple sol-gel route by using inorganic precursors like hafnium (IV) chloride (HfCl$_4$), boric acid (H$_3$BO$_3$) and phenolic resin as a source of hafnium, boron and carbon respectively. The resulting HfB$_2$ powders had an average crystallite size of ~35 nm whilst the HfC powders were ~75 nm in diameter. The precursor gels of HfC and HfB$_2$ were also used to dip coat SiC fibre bundles, on heat treatment, a continuous coating of HfC and HfB$_2$ was obtained. The wettability of the gels was determined using contact angle measurements. The continuity of the coatings on the SiC fibre bundles were characterized using optical and scanning electron microscopy.

INTRODUCTION:

Since the 1960s SiC ceramic matrix materials have been used in thermal protection systems for aerospace vehicles because of their good oxidation and corrosion resistance. They also have a low density, a specific strength greater than that of metals[1], and provide protection at temperatures up to ~1800°C. With the increase in speed in the aerospace industry, however, the use of advanced thermal protection systems (TPS) with greater temperature capability and durability are required to withstand aerothermal heating and its effects, such as thermal shock and thermal ablation that currently limit the performance of hypersonic flights. TPS materials should also be able to retain a significant fraction of their tensile strength and oxidation resistance in environments that could reach up to 3000°C[2]. This higher temperature capability requires additional surface protection either through (i) coating the SiC fibres with durable ultra high temperature ceramics (UHTCs) or (ii) infiltration of UHTCs into the fibre matrix.

HfB$_2$ and HfC are two of the potential candidates for ultra high temperature thermal protection systems in the aerospace industry because of their high melting points and/or better oxidation resistance compared to the other zirconium and tantalum based UHTCs[3]. They also have low coefficient of thermal expansion and high thermal conductivity[4]. These properties have aroused growing interest in these materials, however both ceramics are expensive and are not commercially available with submicron or nano particle sizes. The extensive milling procedures used for particle size reduction of the commercially available microcrystalline powders leads to powder contamination. Both ceramics can be synthesized, either by chemical combination of metal and carbon / boron or by reduction of the metal oxide with carbon / boron. Of these, the latter is the most popular route[3-6]. This approach, however, requires high reaction temperatures of 1900-2300°C, which, in turn, often result in large particle sizes[5]. Thus there is a need to develop a low temperature synthesis procedure for obtaining fine UHTC powders. The present work describes a simple and cheap synthesis process for obtaining phase pure nano-crystalline HfB$_2$ and fine submicron HfC powders using carbothermal/borothermal reduction of a sol-gel precursor. This method uses relatively simple equipment and moderately low reaction temperatures.

*Corresponding Author, Email: B.Vaidhyanathan@lboro.ac.uk, Ph: +441509223152

To synthesize HfC, the conventional carbothermal reduction reaction shown in (1) has been used, whilst a carbo/borothermal reduction reaction, indicated in (2), has been employed to synthesize the HfB$_2$[6].

$$HfO_2 + 3C \longrightarrow HfC + 2CO \qquad (1)$$

$$HfO_2 + B_2O_3 + 5C \longrightarrow HfB_2 + 5CO \qquad (2)$$

As mentioned earlier, coatings of the HfB$_2$ and HfC precursor gels on SiC fibre bundles have been investigated. The synthesized powders can also be used to infiltrate the SiC matrix to improve its high temperature durability.

EXPERIMENTAL PROCEDURE:

Hafnium (IV) chloride (98% pure from Sigma-Aldrich Company Ltd, Dorset, UK) was used as the hafnium source. A phenolic resin was used as the carbon source (Cellobond J2027L, Momentive Speciality Chemicals, Louisville, USA), the char yield was 45%. Boric acid (H$_3$BO$_3$, 99.5% pure) and ethanol (96% pure) were obtained from Fischer Scientific, Loughborough, UK and were used as a source of boron and a solvent respectively. The synthesis of the HfC precursor powder involved the initial dissolution of the hafnium chloride in the ethanol followed by mixing with the phenolic resin in the required amounts. The ratio of Hf to C was kept as 1:3 according to reaction (1). The pH of the mixed solution was adjusted to 5 using hydrochloric acid and the solution mixture was stirred at 60°C for 30 minutes to obtain a gel. The latter was dried at 250°C for an hour and ground to obtain the HfC precursor powder. The approach for the HfB$_2$ was very similar. H$_3$BO$_3$ was dissolved in ethanol at 140°C and mixed with the required amounts of HfCl$_4$ and phenolic resin. The ratio of Hf:B:C were kept as 1:2:5 according to reaction (2). The mixture was stirred at 140°C for half an hour to obtain the gel which was then dried at 250°C and ground to obtain the HfB$_2$ precursor powder.

The carbothermal and carbo/borothermal reduction reactions were performed in a high temperature horizontal tube furnace fitted with a 99.7% pure alumina tube (TSH17/75/450, Elite Thermal Systems Ltd, UK) at 1500°C for 4 hours or at 1600°C for 4 or 5 hours under an argon atmosphere. The heating and cooling rates were maintained at 5°C/min in all cases. An estimation of the particle size and agglomerate size were obtained using transmission electron microscope (TEM 100 CX, JEOL JEM, Munich, Germany) and field emission electron microscopy (FEGSEM 1530 VP, Carl Zeiss (Leo), Oberkochen, Germany) respectively, whilst the crystalline phase of the resulting powders was identified using Cu Kα radiation (Bruker D8 X-Ray Diffractometer, Bruker, Coventry, UK). High temperature X-ray diffraction patterns of these powders were also recorded on heating from 400°C to 1400°C, in air, at 100°C intervals to determine their oxidation behaviour.

The rheological characterization of the HfC and HfB$_2$ precursor gels was carried out at a constant sheer rate using a rheometer (Physica MCR 101, Anton Paar Ltd, Garz, Austria).The HfC and HfB$_2$ precursor gels were subsequently used to manually dip coat Hi-Nicalon SiC fibre bundles. This was carried out by dipping the SiC fibres bundles into the gel, using a pair of forceps and withdrawing them slowly. The gels exhibited good wettability on the fibres. Once coated, the fibres were dried at 250°C and subsequently heat treated at 1600°C for 5 hours. The surfaces of the dried coatings were observed using optical microscopy and the surface of the UHTC coatings was observed using a field emission gun scanning electron microscope.

RESULTS AND DISCUSSION:

Figures 1(a) and 1(b) show the XRD patterns obtained for the HfB$_2$ and HfC precursor powders heat treated at 1500°C for 5 hrs and 1600°C for 4 & 5 hrs. At 1500°C, the formation of HfB$_2$ has been initiated, figure 1(a), and it is complete by 1600°C, 5 hrs. A similar result may be observed for the formation of HfC, figure 1(b). It is also apparent that the intensities of HfC peaks at 1500°C is significantly higher than that of HfB$_2$ peaks, this suggests that HfC has started to form at a lower temperature than HfB$_2$, although both seem to complete formation at the same temperature and time. The patterns have been indexed as hexagonal for HfB$_2$ and face centred cubic for HfC according to the JCPDS card numbers 00-03801398 and 00-03901491 respectively.

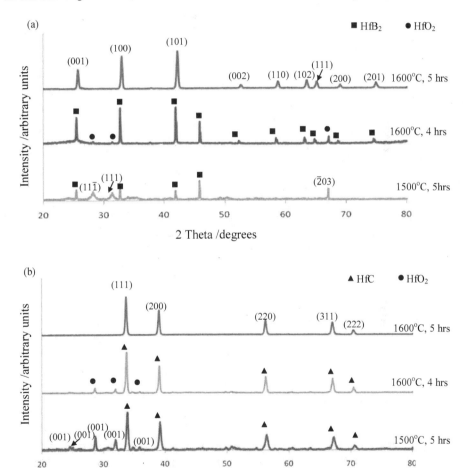

Figure 1: XRD patterns of (a) synthesized HfB$_2$ powder and (b) HfC powder after heat treatment of the precursor gels at 1500 and 1600°C.

Figure 2(a) and (b) shows the TEM images of the synthesized HfB_2 and HfC powders with selected area electron diffraction (SAED) analysis. The crystallite sizes of the synthesized HfB_2 and HfC powders may be seen to exist in the ranges $30 - 40$ nm and $50 - 100$ nm respectively. Both also exhibited agglomeration. The difference in the primary crystallite size may be due to the longer time required from initiation through to the completion for the HfC, as compared to that of HfB_2 (see Figure 1).

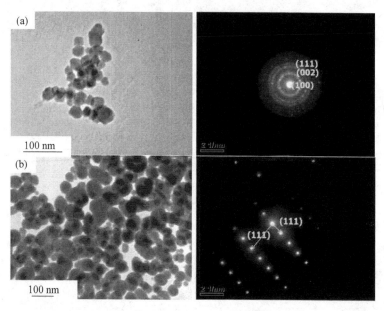

Figure 2: TEM and SAED pattern of synthesized powders (a) HfB_2 and (b) HfC.

However, the FEGSEM pictures; Figures 3(a) and 3(b), of the UHTC powders reveal the presence of micrometer sized agglomerates.

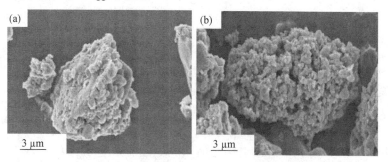

Figure 3: FEGSEM of the synthesized powders (a) HfB_2 and (b) HfC, indicating the presence of agglomerates.

Figures 4(a) and 4(b) provide the high temperature XRD (HTXRD) patterns for the two powders at different temperatures during heating in air; the results shown focus on the temperature regimes where oxidation occurs. From figure 4(a) it is evident that the onset of oxidation for HfB₂ is between 700-800°C, whilst that for HfC, figure 4(b), is between 500-600°C, i.e. at a temperature that is significantly lower than for the HfB₂. This suggests that oxidation resistance of HfB₂ is superior to that of HfC, albeit both completely oxidise below 800°C. This could be due to the differences in the oxidation mechanisms as envisaged in the earlier reports[10], arising from the presence / absence of B₂O₃ phase, in the two UHTC materials investigated. However, more work is needed to elucidate the exact origin of this phenomenon.

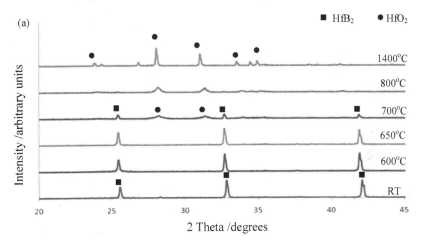

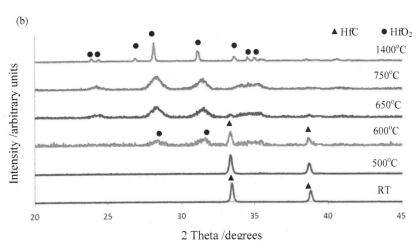

Figure 4: High temperature X-ray diffraction patterns of (a) HfB₂ and (b) HfC powders during heating in air and showing the formation of HfO₂, i.e. the oxidation process.

Comparative high temperature XRD studies on commercial micron sized HfC and HfB$_2$ powders procured from Treibacher Industrie AG, Althofen, Austria also showed a similar trend, with HfC showing a lower oxidation initiation temperature than HfB$_2$. The commercial HfB$_2$ powder, which had an average particle size of 2. 6 μm, oxidized around 700 to 800°C, similar to the much finer sol-gel synthesized powder. Similar results were also noted for the commercial HfC powder, which had an average particle sized of 1. 6 μm, and the sol-gel derived powder, both exhibiting oxidation initiation in the temperature range 500 to 600°C. Though the sol-gel synthesized powders had smaller crystallite size than the micron size commercial powders, they were composed of micron sized agglomerates (see Figure 3). This may have contributed to the similar oxidation behaviour of the different powders.

The viscosity of the HfB$_2$ and HfC precursor gels, used for dip coating the SiC fibre bundles, was found to be 30 mPa s and 20 mPa s respectively after the 30 minute stirring period. The viscosity of the gel increased with the stirring time, probably due to dehydration and skeletal relaxation. Thus by varying the stirring time, the viscosity, and hence the thickness of the coatings could be varied[9]. The sol viscosity showed a monotonous increase at lower temperatures and a rapid increase at higher temperatures representing the cross-linking of the phenolic resin. The viscosity values at different temperatures are listed in Table 1.

Table 1: Variation of viscosity of the HfB$_2$ and HfC gels with temperature

Temperature / °C	Viscosity / Pa.s	
	HfB$_2$	HfC
20	0.03	0.02
60	0.11	0.6
95	0.6	19.4

The optical images of dried coatings on the SiC fibres, shown in Figures 5(a) and (b) do not show any cracks due to drying. The SEM images, Figure 6(a) and (b), of the HfB$_2$ and HfC dip coated SiC fibre bundles also concur with the optical microscopy results.

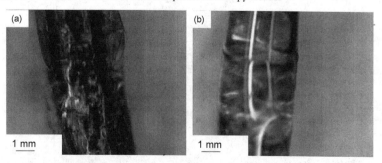

Figure 5: Optical images of SiC fibre bundles dip coated with (a) HfB$_2$ and (b) HfC sols and dried at 250°C.

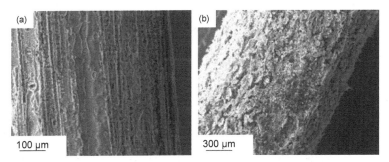

Figure 6: SEM images of SiC fibre bundles dip coated with (a) HfB_2 and (b) HfC sols and heat treated at 1600°C for 5 hrs.

A good wettability on the SiC fibre surface was noticed for both HfB_2 and HfC gels and the contact angles measurements indicated 18° and 8° respectively for them. This is also corroborated in the cross sectional microscopic image of the HfB_2 coated fibres as a representative, Figure 7

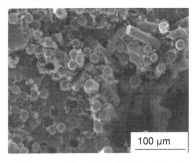

Figure 7: SEM images of fracture surface of the SiC fibre bundles coated with HfB_2 and heat treated at 1600°C for 5 hrs.

The coating thickness was found to increase with the increase in the viscosity and the withdrawal speed as indicated by equation (3)[11].

$$h_o = 0.94(\eta U_o)^{2/3} / v_{LV}^{1/6} (\rho g)^{1/2} \qquad (3)$$

Where: h_o = thickness
η = viscosity
U_o = withdrawal speed
v_{LV} = surface tension
ρ = density
g = acceleration due to gravity

Rather than using a manual dip coating process, an automated process with controlled withdrawal speed is currently being developed to achieve a uniform and thin coating throughout the length of the fibre.

SUMMARY:

Nano-crystalline HfB$_2$ and HfC powders have been synthesized through a simple, cost effective sol-gel route unlike the other expensive CVD routes that require complicated set ups. From the TEM images, the average crystallite size of the HfB$_2$ was found to be 35 nm and that of HfC was found to be 75 nm. From the high temperature XRD results, it can be concluded that HfB$_2$ has a better oxidation resistance than HfC even though the average particle size is smaller for the former than that of the latter. The oxidation performance of the sol-gel prepared UHTC powders were found comparable with that of commercial micro-crystalline powders. Coatings of HfC and HfB$_2$ on SiC fibres showed good adhesion without any cracking. The porosity of the coatings is expected to decrease on reducing the thickness of the coatings[11, 12] and further work is underway.

ACKNOWLEDGEMENT:

The authors thank DSTL, UK for the financial support provided.

REFERENCES:

[1]A. Medkov, P. A. Storozhenko, A. M. T. sirilin, N. I. Steblevskaya, ZrO$_2$ coatings on SiC fibres, *Inorg Mater,* **43**, 162-166, (2007)

[2]S. R. Levine, E. J. Opila, M. C. Halbig, J. D. Kiser, M. Singh, and J. A. Salem, Evaluation of Ultra-High Temperature Ceramics for Aeropropulsion Use, *J. Eur. Ceram. Soc.,* **22**, 2757–67, (2002).

[3]M. D. Sacks, Chang-an Wang, Z. Yang, A. Jain, Carbothermal reduction synthesis of nanocrystalline zirconium carbide and hafnium carbide powders using Solution-derived precursors, *J. Mater. Sci,* **39**, 6057 – 6066, (2004).

[4]J. K. Sonber, T.S.R.Ch. Murthy, C. Subramanian, S. Kumar, R.K. Fotedar, A.K. Suri, Investigations on synthesis of HfB$_2$ and development of a new composite with TiSi^2, *Int. J. Refract. Met. Hard. Mater,* **28**, 201–210, (2010).

[5]Z. Mei, X. Sub, G. Hou, F. Guod and Y. Zhang, Synthesis of HfC powders by carbothermal reduction of modified hafnium alkoxide precursors, *Adv. Mater. Sci,* **148**, 1453-1457, (2011).

[6]D. W. Ni, G. J. Zhang, Y. M. Kan, and P. L. Wangz, Synthesis of Monodispersed Fine Hafnium Diboride Powders Using Carbo/Borothermal Reduction of Hafnium Dioxide, *J. Am. Ceram. Soc.,* **91**, 2709–2712 (2008).

[7]M. Opeka, I. G. Talmy, E. J. Wuchina, J. A. Zaykoski and S. J. Causey, Mechanical, Thermal, and Oxidation Propertiesof Refractory Hafnium and Zirconium Compounds, *J. Eur. Ceram. Soc.,* **19**, 2405-2414, (1999).

[8]R. Savino , M. Fumo, L. Silvestroni, D. Sciti, Arc-jet testing on HfB$_2$ and HfC-based ultra-high temperature ceramic materials, *J. Eur. Ceram. Soc.,* **28**, 1899–1907, (2008).

[9]T. A. Gallo and L. C. Klein, Apparent viscosity of sol-gel processed silica, *J. Non-Cryst. Solids,* **82**, 198-204, (1986).

[10]A. G. Metclafe, N. B. Elsner and D. T. Allen, Opeka, Oxidation of hafnium diboride, "High temperature corrosion and materials chemistry" *Electrochem. Soc. proceedings*, **99-38**, 489-501, (2000).

[11]C. J. Brinker and A. J. Hurd, Fundamentals of sol-gel dip-coating, *J. Phys.*III, **4**, 1231-1242, (1994).

[12]N. I. Baklanova, V. N. Kulyukin, M. A. Korchagin, and N. Z. Lyakhov, Formation of Carbide Coatings on Nicalon Fiber by Gas-Phase Transport Reactions, *J. Mater. Synth. Process*, **6**, (1998).

PRESSURELESS SINTERING AND HOT-PRESSING OF Ti$_2$AlN POWDERS OBTAINED BY SHS PROCESS

L. Chlubny, J. Lis, M.M. Bućko, D. Kata
AGH - University of Science and Technology, Faculty of Material Science and Ceramics, Department of Technology of Ceramics and Refractories, Al. Mickiewicza 30, 30-059, Cracow, Poland

ABSTRACT

In the Ti-Al-N system ternary materials called MAX-phases can be found. These materials are characterised by heterodesmic layer structure, consisting of covalent and metallic chemical bonds. These facts strongly influence their pseudoplastic behavior locating them on the boundary between metals and ceramics, which may lead to many potential applications, for example as a part of ceramic armour. Ti$_2$AlN is one of these nanolaminate materials. To obtain sinterable powders of Ti$_2$AlN of relatively large quantities, Self-propagating High-temperature Synthesis (SHS) was applied. Various substrates such as elementary powders, and titanium and aluminium compounds were used as precursors to produce fine powders of ternary material. Phase compositions of obtained powder were examined by XRD method. After phase analysis, selected powders were densified by pressureless sintering and hot pressing process in various conditions.

INTRODUCTION

Among many advanced ceramic materials such as carbides or nitrides there is a group of ternary compounds referred in literature as H-phases (or Hägg-phases), Nowotny-phases, thermodynamically stable nanolaminates and at last but not least - MAX phases. These compounds have a M$_{n+1}$AX$_n$ stoichiometry, where M is an early transition metal, A is an element of A groups (mostly IIIA or IVA) and X is carbon and/or nitrogen. Heterodesmic structures of these phases are hexagonal, P63/mmc, and specifically layered. They consist of alternate near close-packed layers of M$_6$X octahedrons with strong covalent bonds and layers of A atoms located at the centre of trigonal prisms. The M$_6$X octahedral, similar to those forming respective binary carbides, are connected one to another by shared edges. Variability of chemical composition of the nanolaminates is usually labeled by the symbol describing their stoichiometry, e.g. Ti$_2$AlC represents 211 type phase and Ti$_3$AlC$_2$ – 312 typ. Structurally, differences between the respective phases consist in the number of M layers separating the A-layers: in the 211's there are two whereas in the 321's three M-layers. This new group of compounds was described in details in work of Nowotny et.al, Barsoum and Z. Lin et al. [1-3, 5]. The layered, heterodesmic structure of MAX phases led to an extraordinary set of properties. These materials combine properties of ceramics like high stiffness, moderately low coefficient of thermal expansion and excellent thermal and chemical resistance with low hardness, good compressive strength, high fracture toughness, ductile behavior, good electrical and thermal conductivity characteristic for metals. They can be used to produce ceramic armor based on functionally graded materials (FGM) or as a matrix in ceramic-based composites reinforced by covalent phases.

The SHS is a method that allows obtaining lots of covalent materials such as carbides, borides, nitrides, oxides and intermetallic compounds. The base of this method is utilization of exothermal effect of chemical synthesis, which can proceed in powder bed of solid substrates or as filtration combustion. An external source of heat is used to initiate the process and then the self-sustaining reaction front is propagating through the bed of substrates. This process could be initiated by local ignition or by thermal explosion. The form of synthesized material depends on kind of precursor used for synthesis and technique that was applied. Typical feature of this reaction are low energy consumption, high temperatures obtained during the process, high efficiency and simple apparatus. The lack of control of the process is the disadvantage of this method[4].

The objective of this work was obtaining of fine, sinterable Ti₂AlN powders by SHS method and sintering them by hot-pressing method and pressureless sintering.

PREPARATION

Following the experience gained while synthesizing ternary materials such as Ti₃AlC₂ and also Ti₂AlC [7, 8, 12], as well as earlier experiments with Ti₂AlN synthesis [11, 13, 14], intermetallic materials in the Ti-Al system were used as precursors for synthesis of Ti₂AlN powders. Due to relatively low availability of commercial powders of intermetallic materials from Ti-Al system it was decided to synthesize them by SHS method. In the first stage of the experiment TiAl powder was synthesized by SHS method [7]. Titanium hydride powder, TiH₂, and metallic aluminium powder with grain sizes below 10 m were used as sources of titanium and aluminium. The mixture for SHS had a molar ratio of 1:1 (equations 1).

$$TiH_2 + Al \rightarrow TiAl + H_2 \tag{1}$$

The process of homogenization of powders was conducted in a ball-mill; the time of the process was 12 hours. Afterwards homogenized powders were placed in a graphite crucible which was heated in a graphite furnace in the argon atmosphere up to 1200°C when SHS reaction was initiated. In the next step, products of synthesis were initially crushed in a roll crusher to the grain size ca. 1 mm and then powders were ground in the rotary-vibratory mill for 8 hours in isopropanol, using WC balls as a grinding medium, to the grain size ca. 10 μm[12].

The synthesis of Ti₂AlN was conducted by SHS method with a local ignition system and with use of various precursors. The precursors used for a synthesis were SHS derived TiAl and Ti₃Al powder and commercially available powders of titanium and pure nitrogen. The mixture of precursors for a SHS synthesis was set in appropriate stoichiometric ratio and is presented in equation 2.[13]

$$2\ TiAl + 2\ Ti + N_2 \rightarrow 2\ Ti_2AlN \tag{2}$$

Products of SHS synthesis were ground and the X-ray diffraction analysis method was applied to determine phase composition of the synthesised materials. The basis of phase analysis was data from ICCD [9]. Phase quantities were determined by Rietveld analysis [10].

When the phase composition of synthesized product was confirmed, powders were sintered both by hot-pressing method and pressureless sintering method.

The pressureless sintering was conducted in a graphite furnace in constant nitrogen flow up to the temperature 2200°C. The sintering process was preceded by uniaxial pressing of the selected powder. Afterwards, sintered samples were examined by XRD method to estimate the phase composition of material.

In case of hot-pressing process powders were sintered in the Thermal Technologies 2000 hot press, at temperature range varies from 1100 to 1300°C, annealing time in maximum temperature was 1 hour, sintering was conducted in constant flow of nitrogen. Diameter of obtained samples was 2.5 and 7.6 cm and height of samples was 0.5 cm. The possibility of this approach was proven by Z.J. Lin et al, during hot pressing of TiN + Al + Ti mixture up to 1400 °C under a pressure of 25 MPa as well as by previous researches of authors on the Ti-Al-C-N system [6, 11, 14].

RESULTS AND DISCUSSION

The X-ray diffraction analysis showed that TiAl synthesised by SHS was almost phase pure and contained only about 5% of Ti$_3$Al impurities, as it is presented on Figure 1[11].

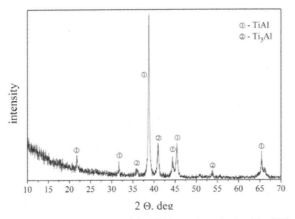

Figure 1. XRD pattern of the TiAl powders obtained by SHS

Also XRD analysis done in case of product of Ti$_2$AlN SHS synthesis, confirmed presence of desired phase. As a result of this reaction mixture of four phases was obtained, where Ti$_2$AlN was the dominating phase (57%). The XRD pattern of obtained powder is presented on Figure 2. The phase composition of SHS product is presented in Table I.

Table I. Products of SHS synthesis of Ti$_2$AlN phase composition

Precursor	Composition, wt %.
TiAl + Ti + N$_2$	57% Ti$_2$AlN, 24% TiN, 11% Ti$_3$Al, 8% Ti$_3$AlN [11]

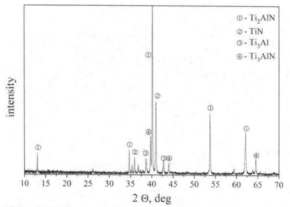

Figure 2. XRD pattern of product of 2 TiAl + 2 Ti + N$_2$ synthesis[11]

Presureless sintering process is presented by sintering curve on Figure 3. It can be observed that sintering process begins at 1600°C; optimal sintering temperature can be estimated at ca. 2070°C. Phase analysis made on sample sintered at the optimal sintering temperature shown transformation of MAX phases into regular TiN like structure (100% of TiN), these results are presented on Fig.4. This result corresponds with observations that were made for other MAX phases materials such as Ti$_2$AlC, Ti$_3$AlC$_2$ and Ti$_3$SiC$_2$.

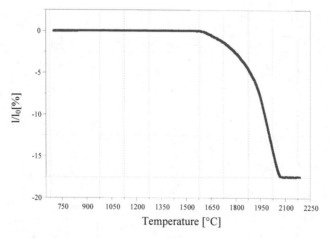

Figure 3. Pressureless sintering curve of SHS derived Ti$_2$AlN.

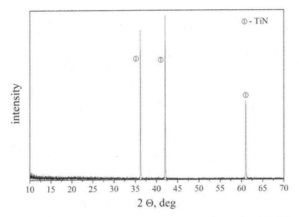

Figure 4. XRD pattern of Ti₂AlN powder sintered at 2070°C.

In case of hot pressed sample, it was observed that amount of MAX phase is increasing with the increasing temperature of sintering process. In 1300°C dense, polycrystalline material, containing over 75% of Ti₂AlN was obtained. The XRD results are presented on Figures 5-7 and detailed phase composition of hot-pressed samples are presented in Table II. Above 1300°C decomposition of MAX phases takes place and hexagonal structure changes into regular structure. This phenomena seems to be characteristic for ternary nanolaminate materials and can be also observed in case of materials in Ti-Al-C and Ti-Si-C systems.

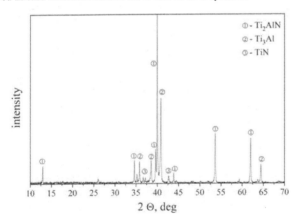

Figure 5. XRD pattern of Ti₂AlN hot pressed at 1100°C.

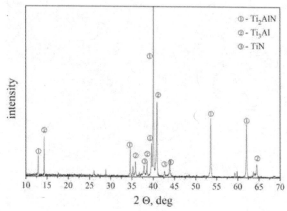

Figure 6. XRD pattern of Ti₂AlN hot pressed at 1200°C.

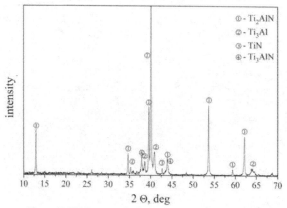

Figure 7. XRD pattern of Ti₂AlN hot pressed at 1300°C.

Table II. Phase composition of hot pressed Ti₂AlN powders.

Temperature [°C]	Composition, wt %.
1100	61.3% Ti₂AlN, 4.8% TiN, 33.9% Ti₃Al
1200	72.7% Ti₂AlN, 2.2% TiN, 25.1% Ti₃Al
1300	75.8% Ti₂AlN, 1.3% TiN, 13.4% Ti₃Al, 9.6 Ti₃AlN

CONCLUSIONS

Obtaining of sinterable Ti$_2$AlN powders by Self-propagating High-temperature Synthesis (SHS) is possible. This method proved to be an effective, efficient and economic. Powders are characterized by relatively high amount of ternary nanolaminate materials (57% of Ti$_2$AlN in the sample volume). For SHS synthesis of this MAX phase, intermetallic compound namely TiAl had to be used as a precursor.

Sintering of SHS derived Ti$_2$AlN powders is possible by hot-pressing method. During the process further chemical reaction are observed. This reactions lead to significant increase of amount of Ti$_2$AlN in the sample volume. Optimal sintering temperature was estimated at 1300°C in case of HP process. Above this temperature decomposition of MAX phase is observed. This fact closely corresponds with results obtained by Z.J. Lin et al. during the hot pressing of TiN + Ti + Al mixture, where optimal process temperature was establish on 1400°C.[6]

Changes of phase composition are observed during pressureless sintering process as it was observed in case of other SHS derived MAX phases such as Ti$_3$SiC$_2$, Ti$_2$AlC and Ti$_3$AlC$_2$. In this case it was not possible to establish optimal sintering temperature. It was observed that hexagonal structure of MAX phase was transformed into regular TiN structure. Basing on these observations it can be concluded that HP process is more adequate for sintering of SHS derived Ti$_2$AlN powders.

Mechanical properties of dense, polycrystalline samples will be examined to confirm their pseudoplastic features.

ACKNOWLEDGMENTS

This work was supported by the Polish Ministry of Science and Higher Education under the grant no. POIG.01.01.02-00-097/09.

REFERENCES

[1] W. Jeitschko, H. Nowotny, F.Benesovsky, Kohlenstoffhaltige ternare Verbindungen (H-Phase). *Monatsh. Chem.* **94**, 1963, p 672-678

[2] H. Nowotny, Structurchemie Einiger Verbindungen der Ubergangsmetalle mit den Elementen C, Si, Ge, Sn. *Prog. Solid State Chem.* **2** 1970, p 27

[3] M.W. Barsoum: The MN+1AXN Phases a New Class of Solids; Thermodynamically Stable Nanolaminates- *Prog Solid St. Chem.* **28**, 2000, p 201-281

[4] J.Lis: Spiekalne proszki związków kowalencyjnych otrzymywane metodą Samorozwijającej się Syntezy Wysokotemperaturowej (SHS) - *Ceramics* **44** : (1994) (*in Polish*)

[5] Z. Lin, M. Li, Y. Zhou: TEM Investigations on Layered Ternary Ceramics; *J Mater Sci Technol* **23** (2007) 145-165

[6] Z.J. Lin, M.J. Zhuo, M.S. Li, J.Y. Wanga, Y.C. Zhou: Synthesis and microstructure of layered-ternary Ti$_2$AlN ceramic; Scripta Mater **56** (2007) 1115–1118

[7] L. Chlubny, M.M. Bucko, J. Lis "Intermetalics as a precursors in SHS synthesis of the materials in Ti-Al-C-N system" *Advances in Science and Technology*, **45**, 2006, p 1047-1051

[8] L. Chlubny, M.M. Bucko, J. Lis "Phase Evolution and Properties of Ti$_2$AlN Based Materials, Obtained by SHS Method" Mechanical Properties and Processing of Ceramic Binary, Ternary and Composite Systems, *Ceramic Engineering and Science Proceedings*, Volume **29**, Issue 2, 2008, Jonathan Salem, Greg Hilmas, and William Fahrenholtz, editors; Tatsuki Ohji and Andrew Wereszczak, volume editors, 2008, p 13-20

[9] "Joint Commitee for Powder Diffraction Standards: International Center for Diffraction Data"

[10] H. M. Rietveld: "A profile refinement method for nuclear and magnetic structures." J. Appl. Cryst. **2** (1969) p. 65-71

[11] L. Chlubny: New materials in Ti-Al-C-N system. - PhD Thesis. AGH-University of Science and Technology, Kraków 2006. (*in Polish*)

[12] L. Chlubny, J. Lis, M.M. Bucko: Preparation of Ti$_3$AlC$_2$ and Ti$_2$AlC powders by SHS method MS&T Pittsburgh 09: Material Science and Technology 2009, 2009, p 2205-2213

[13] L. Chlubny, J. Lis, M.M. Bucko "Titanium and Aluminium Based Compounds as a Precursors for SHS of Ti$_2$AlN" Nanolaminated Ternary Carbides and Nitrides, *Ceramic Engineering and Science Proceedings*, Volume **31**, Issue 10, 2010, Sanjay Mathur and Tatsuki Ohji, volume editors, 2010, p 153-160

[14] L. Chlubny, J. Lis, M. M. Bucko "Influence of Sintering Conditions on Phase Evolution of SHS-Derived Materials in the Ti-Al-N System" ECERS 2007, *Proceedings of the 10th international conference of the European Ceramic Society* : June 17–21, 2007 Berlin. eds. J. G. Heinrich, C. G. Aneziris. Baden-Baden, Göller Verlag GmbH, 2007, p. 1155–1158.

DIFFRACTION STUDY OF SELF-RECOVERY IN DECOMPOSED Al$_2$TiO$_5$ DURING VACUUM ANNEALING

I.M. Low and W.K. Pang
Centre for Materials Research, Department of Imaging & Applied Physics, Curtin University of Technology, GPO Box U1987, Perth, WA 6845, Australia

ABSTRACT

The ability of decomposed Al$_2$TiO$_5$ to undergo self-recovery or reformation during vacuum annealing was characterised by in-situ neutron diffraction. It is shown that the process of phase decomposition in Al$_2$TiO$_5$ was reversible and that reformation occurred readily when decomposed Al$_2$TiO$_5$ was re-heated above 1300°C. The kinetics of isothermal and temperature-dependent self-recovery was modelled using the Avrami equation. The influence of grain-size on the Avramic kinetics of self-recovery was also evident.

INTRODUCTION

Aluminium titanate (Al$_2$TiO$_5$) is an excellent refractory and thermal shock resistant material due to its relatively low thermal expansion coefficient and high melting point. It is one of several materials which is isomorphous with the mineral pseudobrookite (Fe$_2$TiO$_5$).[1,2] In this structure, each Al^{3+} or Ti^{4+} cation is surrounded by six oxygen ions forming distorted oxygen octahedra. These AlO$_6$ or TiO$_6$ octahedra form (001) oriented double chains weakly bonded by shared edges. This structural feature is responsible for the strong thermal expansion anisotropy which generates localised internal stresses to cause severe microcracking. Although this microcracking weakens the material, it imparts a desirable low thermal expansion coefficient and an excellent thermal shock resistance. However, at elevated temperature, Al$_2$TiO$_5$ is only thermodynamically stable above 1280°C and undergoes a eutectoid-like decomposition to α-Al$_2$O$_3$ and TiO$_2$ (rutile) within the temperature range 900-1280°C.[3-13] This undesirable decomposition has limited its wider application.

In our recent studies,[12-15] microstructure and furnace atmosphere have been observed to have a profound influence on the thermal stability of Al$_2$TiO$_5$. For instance, the decomposition rate of Al$_2$TiO$_5$ at 1100°C was significantly enhanced in vacuum (10^{-4} torr) or argon where >90% of Al$_2$TiO$_5$ decomposed after only 4 h annealing when compared to less than 10% in atmospheric air.[12,13] This suggests that the process of decomposition of Al$_2$TiO$_5$ is susceptible to environmental attack or sensitive to the variations in the oxygen partial pressure during ageing. The stark contrast in the mechanism of phase decomposition is believed to arise from the vast differences in the oxygen partial pressure that exists between air and vacuum.

A similar phenomenon has been observed for Al$_2$TiO$_5$ samples having different grain sizes whereby the rate of phase decomposition increased as the grain size decreased.[14] The reason for this grain-size effect is unclear at this stage although the kinetics of decomposition may be rate limited by processes occurring at the grain boundaries. It has also been shown that the process of decomposition in Al$_2$TiO$_5$ is reversible whereby self-recovery occurs readily when decomposed Al$_2$TiO$_5$ is re-heated above 1300°C.[15,16]

In this paper, we present results on the role of grain size on the capability of previously decomposed Al$_2$TiO$_5$ to self-recover in vacuum when it is reheated above 1300°C. The temperature-dependent thermal stability and the capacity of Al$_2$TiO$_5$ to self-recover have been characterized using high-temperature neutron diffraction to study the phase changes in real time.

169

EXPERIMENTAL METHODS
Sample Preparation

The starting powders used for the synthesis of Al_2TiO_5 (AT) consisted of high purity commercial alumina (99.9% Al_2O_3) and rutile (99.5% TiO_2). One mole of alumina powder and one mole of rutile powder were initially mixed using a mortar and pestle. The powder mixture was then wet mixed in ethanol using a Turbula mixer for 2.0 h. The slurry was then dried in a ventilated oven at 100 C for 24 h. The dried powder was uniaxially-pressed in a steel die at 150 MPa to form cylindrical bars of length 20 mm and diameter 15 mm, followed by sintering in a air-ventilated furnace at (a) 1400°C in air for 1 h to achieve a fine-grained microstructure (-1-$3\mu m$), and (b) 1600°C in air for 4 h to achieve coarse-grained (-20-$30\mu m$) Al_2TiO_5. The samples were then completely decomposed by annealing them in an air-ventilated furnace at 1100°C for 10 h. Figure 1 shows the typical diffraction patterns of samples before and after decomposition.

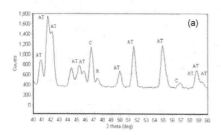
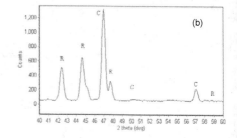

Fig. 1: Typical neutron diffraction plots of Al_2TiO_5 (a) before and (b) after decomposition 1100 C in air. [Legend: AT = Al_2TiO_5; C = corundum; R = rutile]

In-Situ Neutron Diffraction (ND)

In-situ neutron diffraction was used to monitor the structural evolution of self-recovery in previously decomposed Al_2TiO_5 at high-temperature in real time. Diffraction patterns were collected using the Polaris – the high intensity, medium resolution powder diffractometer at the UK pulsed spallation neutron source ISIS, Rutherford Appleton Laboratory. Sample was held in a basket made from thin tantalum wire and mounted in a Risø-design high-temperature furnace (Risø National Laboratory, Roskilde). Fitted with a thin tantalum foil element and tantalum and vanadium heat shields, this furnace is capable of reaching 2000 °C and operates under a high dynamic (i.e. continuously pumped) vacuum (pressure $< 7.5\times10^{-6}$ Torr). Temperature monitoring and control was achieved using type W5 thermocouples connected to Eurotherm 3504 controllers. Collimating slits (manufactured from neutron-absorbing boron nitride) mounted on the furnace in the scattered beam direction enable diffraction patterns free from Bragg reflections off the tantalum element and heat shields to be collected in the Polaris $2\theta =90°$ detectors. A reference diffraction pattern was collected at room temperature while the furnace was initially evacuated, then Al_2TiO_5 samples were heated rapidly up from room temperature to 1450°C for the coarse-grained sample and 1500°C for the fine grained sample. Figure 2 shows the heating protocols used during the experiment for each sample. The relative abundance of phases present was computed using the Rietveld method. The models used to calculate the phase abundance were Maslen et al.[17] for alumina, Epicier et al.[18] for Al_2TiO_5, and Howard et al.[19] for rutile. The software used to analyze the data was Fullprof-WinPlotR.

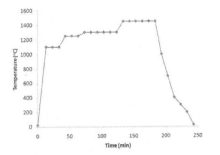
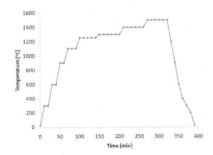

RESULTS AND DISCUSSION
Effect of grain size on Self-Recovery

Figure 2 shows the capability of coarse-grained decomposed Al$_2$TiO$_5$ to self-recover when it was reheated from room temperature to 1450 C. It is clearly shown that self-recovery takes place at ~1450°C through the rapid reaction of corundum and rutile to form Al$_2$TiO$_5$ but with <40 wt% phase purity. In contrast, >65 wt% Al$_2$TiO$_5$ reformed for the fine-grained sample and the process of self-recovery commenced at a lower temperature of ~1400°C (Fig. 3). It appears that fine grains impart a greater rate of atomic diffusion along the grain boundaries to promote the reformation of Al$_2$TiO$_5$ from reaction between Al$_2$O$_3$ and TiO$_2$ as follows:

$$Al_2O_3 \; + \; TiO_2 \quad \leftrightarrow \quad Al_2TiO_5 \qquad (1)$$

To the best of our knowledge, this is the first time that grain size has been shown to affect the propensity of self-recovery in Al$_2$TiO$_5$. This capability of self-recovery further suggests the process of self-reformation is diffusion-controlled and reversible as indicated in Equation (1). However, it should be noted here that the ability of decomposed Al$_2$TiO$_5$ to self-recover in vacuum is much inferior when compared to near-complete self-recovering when annealed in air.[15,16] The underlying reason for this difference is unclear although the role of oxygen partial pressure in the furnace may play a vital role. Nonetheless, the implication of this phenomenon is far-reaching whereby it may be possible to restore the decomposed Al$_2$TiO$_5$ to its original condition by thermal annealing in air at >1400°C.

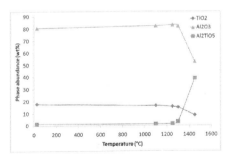
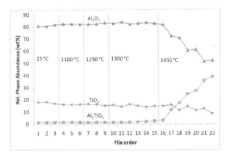

Fig. 2: The propensity of coarse-grained Al$_2$TiO$_5$ to self-recover during reheating to 1450 C in vacuum as a function of: (a) temperature, and (b) time & temperature.

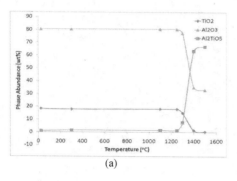

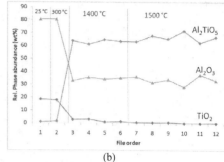

(a)

(b)

Fig. 3: The propensity of fine-grained Al_2TiO_5 to self-recover during reheating to 1500 C in vacuum as a function of: (a) temperature, and (b) time & temperature.

Avrami Kinetics of Self-Recovery

The role of gain size on the kinetics of self-recovery during vacuum-annealing of decomposed Al_2TiO_5 at elevated temperature was modelled using the Avrami equation (see Fig. 4). The Avrami rate constant (k) and Avrami exponent (n) were determined to be 8.15×10^{-15} min^{-n} and 5.88 for coarse-grained and 5.62×10^{-7} min^{-n} and 2.50 for fine-grained.

The isothermal self-recovery of coarse-grained Al_2TiO_5 during vacuum-annealing at 1450°C was fitted with both the least-squares linear regression and the modified Avrami equation (see Fig. 5) The corresponding least-squares regression exponent (R^2) and Avrami constants (k and n) of isothermal decomposition were determined to be 0.96, 6.23×10^{-5} min^{-n} and 2.02 respectively. This suggests that the kinetics of Al_2TiO_5 self-recovery is better described by an exponential Avrami rate reaction than a linear rate reaction.

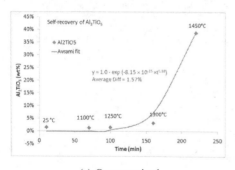

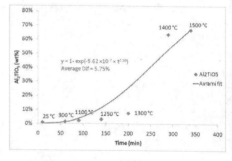

(a) Coarse-grained

(b) Fine-grained

Fig. 4: Effect of (a) coarse-grain and (b) fine-grain on the Avrami fit of self-recovery in vacuum-annealed Al_2TiO_5 at elevated temperature.

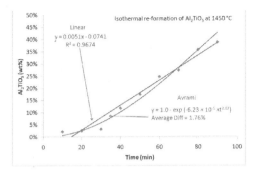

Fig. 5: Least-squares linear regression and Avrami fits of self-recovery during isothermal vacuum-annealing of coarse-grained Al₂TiO₅ at 1450 °C

Microstructural Features

Fig. 6 (a) shows the typical microstructures of as-decomposed fine-grained Al₂TiO₅ prior to vacuum-annealing. The existence of corundum (grey grains) and rutile (white grains) in the microstructure is evident. Following vacuum-annealing at 1500°C for 0.5 h, self-recovery occurred with the formation of Al₂TiO₅ grains. A similar process of self-recovery also occurred for the coarse-grained sample vacuum-annealed at 1450°C for 0.5 h (Fig. 6c). As the microstructure becomes coarser, the degree of self-recovery appears to become less and is least for the coarse-grained sample. This observation is consistent with the neutron diffraction results shown in Figs. 2 & 3 above.

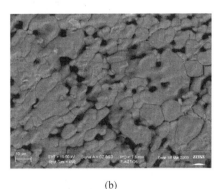

(b)

(a)

(c)

Fig. 6: Scanning electron micrographs showing the microstructures of (a) decomposed fine-grained Al$_2$TiO$_5$, (b) self-recovered fine-grained Al$_2$TiO$_5$, and (c) self-recovered coarse-grained Al$_2$TiO$_5$.

CONCLUSIONS

The effect of grain size on the capacity of Al$_2$TiO$_5$ to undergo self-recovery in the temperature range 20-1500°C was dynamically examined by neutron diffraction. The ability of Al$_2$TiO$_5$ to self-recover in vacuum increased as the grain size decreased probably through an enhanced atomic diffusion process by virtue of increased surface area and grain boundaries. The Avrami kinetics of Al$_2$TiO$_5$ self-recovery was also dependent on grain size.

ACKNOWLEDGEMENTS

This work was supported by a LIEF grant (LE0882725) and the collection of neutron diffraction data was conducted at ISIS (RB1010064) which was provided by the Science and Technology Facilities Council, with financial support from an AMRFP grant. We thank Dr. R. Smith of ISIS and Prof. E. Wu of IMR for assistance with the data collection.

REFERENCES
[1]A.E. Austin and C.M. Schwartz, The Crystal Structure of Aluminium Titanate, *Acta Cryst.* **6**, 812-13 (1953).

[2]B. Morosin and R.W. Lynch, Structure Studies on Al$_2$TiO$_5$ at Room Temperature and at 600°C, *Acta Cryst.* B. **28**, 1040-1046 (1972).

[3]H.A.J. Thomas and R. Stevens, Aluminium Ttitanate – a Literature Review. Part 1: Microcracking Phenomena, *Br. Ceram Trans. J.* **88**, 144-90 (1989).

[4]H.A.J. Thomas and R. Stevens, Aluminium Titanate - A literature Review. Part 2: Engineering Properties and Thermal Stability, *Br. Ceram Trans. J.* **88**, 184-190 (1989).

[5]V. Buscaglia, P. Nanni, G. Battilana, G. Aliprandi, and C. Carry, Reaction Sintering of Aluminium Titanate: 1 - Effect of MgO Addition, *J. Eur. Ceram. Soc.* **13**, 411-417 (1994)

[6]G. Tilloca, Thermal Stabilization of Aluminium Titanate and Properties of Aluminium Titanate Solid Solutions, *J. Mater. Sci.* **26**, 2809-2814 (1991).

[7]E., Kato, K. Daimon and Y. Kobayashi, Factors Affecting Decomposition Temperature of β- Al$_2$TiO$_5$, *J. Am. Ceram. Soc.* **63**, 355-356 (1980).

[8]R.W. Grimes and J. Pilling, Defect Formation in -Al$_2$TiO$_5$ and its Influence on Structure Stability, *J. Mater. Sci.* **29**, 2245-49 (1994).

[9]M. Ishitsuka, T. Sato, T. Endo and M. Shimada, Synthesis and Thermal Stability of Aluminium Titanate Solid Solutions, *J. Am. Ceram. Soc.* **70**, 69-71 (1987).

[10]B. Freudenberg and A. Mocellin, Aluminum Titanate Formation by Solid-State Reaction of Coarse Al$_2$O$_3$ and TiO$_2$ Powders, *J. Am. Ceram. Soc.* **71**, 22-28 (1988).

[11]B. Freudenberg and A. Mocellin, Aluminium Titanate Formation by Solid State Reaction of Al$_2$O$_3$ and TiO$_2$ Single Crystals, *J. Mater. Sci.* **25**, 3701-3708 (1990).

[12]I.M. Low, D. Lawrence, and R.I. Smith, Factors Controlling the Thermal Stability of Aluminium Titanate in Vacuum, *J. Am. Ceram. Soc.* **88**, 2957-2961 (2005).

[13]I.M. Low, Z. Oo and B. O'Connor, Effect of Atmospheres on the Thermal Stability of Aluminium Titanate, *Physica B: Condensed Matter*, **385-386**, 502-504, (2006).

[14]I.M. Low and Z. Oo, Effect of Grain-Size and Atmosphere on the Thermal Stability of Aluminium Titanate, *AIP Conf. Proc.* **1202**, 27-31 (2010).

[15]I.M. Low and Z. Oo, Reformation of Phase Composition in Decomposed Aluminium Titanate, *Mater. Chem. & Phys.* **111**, 9-12 (2008).

[16]I.M. Low and Z. Oo, In-Situ Diffraction Study of Self-Recovery in aluminium titanate, *J. Am. Ceram. Soc.* **91**, 1027-1029 (2008).

[17]E.N. Maslen, V.A. Streltsov, N.R. Streltsova, N. Ishizawa and Y. Satow, Synchrotron X-Ray Study of the Electron Density in -Al$_2$O$_3$, *Acta Crystallographica*, **B49**, 937-980 (1993).

[18]T. Epicier, G. Thomas, H. Wohlfromm and J.S. Moya, High Resolution Electron Microscopy Study of the Cationic Disorder in Al$_2$TiO$_5$, *J. Mater. Res.* **6**, 138-145 (1991).

[19]C.J. Howard, T.M. Sabine and F. Dickson, Structural and Thermal Parameters for Rutile and Anatase, *Acta Cryst. B*, **47**, 462-468 (1991).

KINETICS OF PHASE DECOMPOSITION IN Ti$_4$AlN$_3$ AND Ti$_2$AlN - A COMPARATIVE DIFFRACTION STUDY

W.K. Pang[1], R.I. Smith[2], V.K. Peterson[3], I.M. Low[1]

[1]Centre for Materials Research, Curtin University of Technology, GPO Box U1987, Perth, WA 6845.
[2]ISIS Facility, Science and Technology Facilities Council, Rutherford Appleton Laboratory, Harwell Science and Innovation Campus, Didcot, Oxfordshire OX11 0QX, UK.
[3]The Bragg Institute, Australian Nuclear Science and Technology Organisation, Sydney, NSW 2234, Australia.

ABSTRACT

A method for determining the kinetics of and the activation energy for the thermal dissociation of Ti$_2$AlN and Ti$_4$AlN$_3$ in vacuum using *in-situ* time-of-flight (ToF) neutron diffraction is described. We discuss the thermal stability and phase transitions in Ti$_2$AlN and Ti$_4$AlN$_3$. The 4th order polynomial function of time $[y(t)=At^4+Bt^3+Ct^2+Dt+E]$ is used to describe the isothermal decomposition and to calculate the mean value of the reaction rate constant at 1400, 1450, 1500, and 1550 °C. The rate constants at these four temperatures are used to determine the activation energy for the decomposition of Ti$_4$AlN$_3$ and Ti$_2$AlN in vacuum using the Arrhenius equation, which are found to be 410.8 ± 50.0 and 573.8 ± 130 kJ/mol, respectively.

INTRODUCTION

Ternary nitrides, Ti$_4$AlN$_3$ and Ti$_2$AlN, are members of a group of novel layered ceramic materials, *MAX* phases, which show mixed metallic and ceramic properties, e.g., a low density, low thermal expansion coefficient, high modulus, high strength, high-temperature oxidation resistance, good electrical and thermal conduction, good tolerance to physical damage, resistance to thermal shock, as well as being relatively easy to machine.[1-8]

There is interest in the research community concerning the energy required to initiae the solid-state reaction and its corresponding kinetic mechanisms. Many solid state reactions can be represented by the equation $F(a) = kt$, where a is the fraction of material transformed in time (t), and k is the reaction rate constant.[9-10] The kinetic regimes of interest in these reactions include (i) diffusion control reactions, (ii) phase boundary controlled reactions (first-order reactions), and (iii) reactions that can be described by the Avrami-Erofe'ev equations. In all these reactions, k (the reaction rate constant) can be readily determined and used to calculate the activation energy using Arrhenius' equation.[11]

In this paper, we describe the use of ToF neutron diffraction for the study of the kinetics of reactions leading to phase instability of Ti$_4$AlN$_3$ and Ti$_2$AlN in high-vacuum at a range of temperatures up to 1600 °C. A new method combining the use of 4th order polynomial functions that describe the isothermal decomposition of Ti$_4$AlN$_3$ at 1400, 1450, 1500, and 1550 °C, with the Arrhenius equation is applied to determine the activation energies for these processes. The phase transitions during Ti$_4$AlN$_3$ and Ti$_2$AlN decomposition in vacuum up to 1600 °C are also reported.

EXPERIMENTAL PROCEDURE

Dense, hot-pressed cylindrical Ti$_4$AlN$_3$ and Ti$_2$AlN samples of 15 mm diameter and 25 mm height were used in this study. The samples were prepared by hot-pressing of powdered mixtures of TiN (99.3% purity, 2.03μm), Ti (99.0% purity, 2.48 μm) and Al (99.8% purity, 1.50 μm). The powders were mixed in the molar ratio of Ti: Al:TiN = 1:1:3. The mixture was initially mixed in ethanol for 24 h, and then hot-pressed in an Argon atmosphere at a heating rate of 50 °C/min until 1400 °C, where it was held for 2 h. The pressure during the hot-pressing was 30 MPa. Cylindrical Ti$_2$AlN samples were prepared by the same method but using powder mixtures of TiN (99.3%

purity, 2.03 μm), Ti (99.0% purity, 2.48 μm) and Al (99.8% purity, 1.50 μm) in the molar ratio of Ti:Al:TiN = 1:1:1. The Ti_4AlN_3 sample was found to contain 9.79 wt.% TiN, 1.71 wt.% Ti_2AlN, and had ≤ 0.5% porosity.

In-situ ToF powder neutron diffraction was used to monitor the structural evolution of the phase decomposition of Ti_4AlN_3 and Ti_2AlN at high-temperature in real time. Diffraction patterns were collected using Polaris, the high-intensity medium-resolution powder diffractometer, at the UK pulsed spallation neutron source ISIS, Rutherford Appleton Laboratory. Samples were held in a thin tantalum wire basket and mounted in a high-temperature furnace (Risø National Laboratory, Roskilde). The furnace was fitted with a thin tantalum foil element and tantalum and vanadium heat shields that allows it to reach 2000 °C, and operates under a high dynamic, i.e., it is continuously evacuated with a vacuum < 7.5×10^{-5} Torr. Temperature monitoring and control was achieved using type W5 thermocouples and Eurotherm 3504 controllers. Neutron-absorbing boron nitride was mounted on the furnace in the scattered beam direction A mask made from neutron- absorbing boron nitride was mounted on the furnace in the scattered beam direction to prevent incident neutrons scattered by the tantalum element and heat shields from reaching the Polaris 2θ = 90° detectors. A precision electronic scale (reading to five decimal places) was used to weigh the sample before it was loaded into the furnace.

A reference diffraction pattern was collected at room temperature while the furnace was initially evacuated, then Ti_4AlN_3 and Ti_2AlN samples were heated rapidly accordingly to the heating protocol shown in Figure 1.

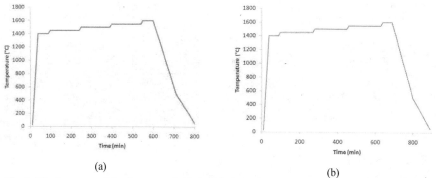

(a) (b)

Fig. 1: Heating protocol for the *in-situ* ToF neutron diffraction experiment for (a) Ti_4AlN_3, and (b) Ti_2AlN.

The sample was held isothermally for an extended period at chosen temperatures during which a series of diffraction patterns, each of 10 min duration, were collected. After the last measurement at 1600 °C, the furnace was cooled to room temperature, and the sample carefully removed from the furnace and weighed to determine the mass of Ti and Al lost through evaporation.

Normalised data from the Polaris 90° detector bank over the *d*-spacing range of 0.32–4.20 Å were analysed using Rietveld analysis with Fullprof-WinPlotR[12] to allow derivation of compositional changes, i.e., Ti_4AlN_3, Ti_2AlN, and TiN contents, during vacuum annealing. At each isothermal hold, the abundance of the ternary nitride, Ti_4AlN_3 or Ti_2AlN, in wt.% was plotted as a function of time and fitted with a 4th order polynomial function. The derivative of the polynomial function was used to obtain a function describing the reaction rate, $[y'(t)]$. A mean value of the function over the isothermal period can be treated as an average reaction rate for the isothermal decomposition at the specific temperature. The calculated average reaction rates for each temperature were used to determine the activation energy for the decomposition process.

Scanning electron micrographs of as-received and vacuum-annealed Ti_4AlN_3 and Ti_2AlN samples were acquired using the Zeiss (Oberkochen, Germany) EVO 40XVP SEM with an accelerating voltage of 15 keV. The samples were not gold or carbon-coated before the microstructure examination and the images were taken using secondary electrons.

RESULTS AND DISCUSSION
Phase Transition
The residual values of refinements, R_{wp} and R_{exp}, ranged from 14.3 to 20.4 and 7.46 to 13.6, respectively. The goodness-of fit (GOF) ranged from 1.70 to 2.48. Typical Rietveld profile fits are shown in Fig. 2 for Ti_4AlN_3 and Fig. 3 for Ti_2AlN, before and after vacuum annealing.

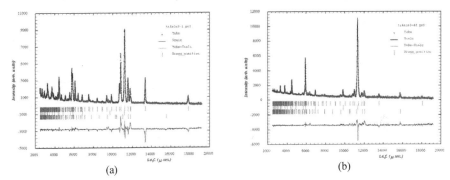

(a)   (b)

Fig. 2: Rietveld profile fits for Ti_4AlN_3 (a) before and (b) after vacuum decomposition. The difference between the measured and calculated patterns are shown as the blue line and vertical bars are the reflection markers for Ti_4AlN_3 (top), TiN_x (middle) and Ti_2AlN (bottom).

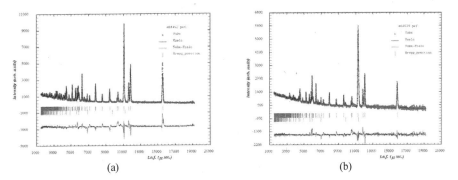

(a)   (b)

Fig. 3: Rietveld profile fits for Ti_2AlN (a) before and (b) after vacuum decomposition. Measured data are black crosses and the calculated pattern is the red solid line. The difference between the measured and calculated patterns are shown as the blue line and vertical bars are the reflection markers for Ti_2AlN (top) and TiN_x (bottom).

The change in sample composition occurring during the vacuum decomposition of Ti_4AlN_3 up to 1600 °C are shown in Fig. 4a. At high temperature (> 1400 °C), Ti_4AlN_3 decomposed to TiN_x *via* sublimation of Ti and Al. During the first isothermal hold (1400 °C) for 60 min, the amount of

Ti_4AlN_3 remained almost constant. At 1450 °C, the relative amount of Ti_4AlN_3 decreased by 4% after 150 mins. At 1500 °C, this amount dropped by a further 15% after 150 mins, and at 1550 °C a further 71 % decrease (to 10 %) was observed after 150 mins. At 1600 °C after 10 mins, almost all the remaining Ti_4AlN_3 had disappeared, indicating almost complete decomposition into TiN_x. During both the 60 minute hold at 1600 °C and subsequent cooling to room temperature, there were no further phase changes detected. On the other hand, the abundance of Ti_2AlN in this sample remained constant during the whole heating and cooling process. These results indicate that Ti_4AlN_3 is readily decomposed into TiN_x under vacuum and that the decomposition rate increases with temperature, but that Ti_2AlN is relatively more stable at these temperatures.

During vacuum annealing Ti_2AlN also decomposed into TiN_x, but these results suggest that Ti_2AlN exhibits a much better resistance against vacuum decomposition than Ti_4AlN_3 (see Fig. 4b). After holding at 1400 °C for 60 min, followed by 1450 °C for 180 min, and 1500 °C for 180 min, only a small decrease in the relative phase abundance of Ti_2AlN, < 1%, was observed. Similarly, only a 2.5% drop in the relative phase abundance of Ti_2AlN was observed after the 180 min hold at 1550 °C. The rate of Ti_2AlN decomposition was accelerated at 1600 °C, where a 10% decrease in the relative phase abundance of Ti_2AlN was observed after 60 min. These results indicate that the decomposition of Ti_2AlN starts at temperatures above ~1450 °C and that the decomposition rate increases dramatically above 1550 °C.

The high-resolution neutron powder diffractometer (HRPD), Echidna, at the Bragg Institute in Australia was used to determine the relative phase abundances of the vacuum-decomposed Ti_4AlN_3 and Ti_2AlN, and indicated the former to contain 99.0 wt% TiN_x and 1.0 wt% Ti_2AlN (Fig. 5a) and the latter to contain 10.8 wt% TiN_x, 89.2 wt% Ti_2AlN (Fig. 5b).

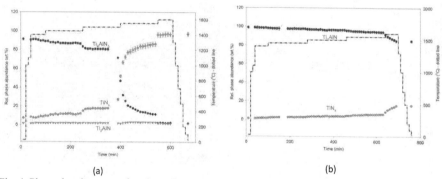

(a)   (b)

Fig. 4: Phase abundance as a function of temperature and time for decomposition of (a) Ti_4AlN_3 and (b) Ti_2AlN in vacuum.

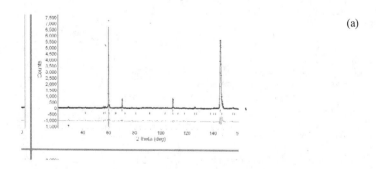

(a)

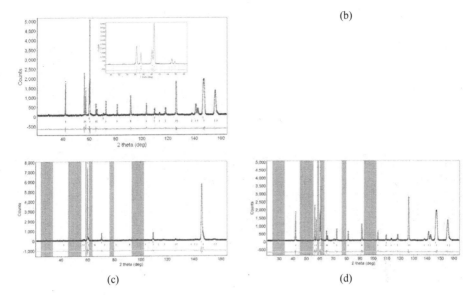

Fig. 5: The Rietveld profile fit of high-resolution diffraction data of vacuum-decomposed samples at 1600 °C; (a) Ti$_4$AlN$_3$ (χ^2 = 6.5, R$_{wp}$ = 15.4, R$_{exp}$ = 6.0), and (b) Ti$_2$AlN (χ^2 = 4.0, R$_{wp}$ = 14.16, R$_{exp}$ = 7.03), (c) Ti$_4$AlN$_3$ (χ^2 = 6.67, R$_{wp}$ = 14.4, R$_{exp}$ = 5.6), and (d) Ti$_2$AlN (χ^2 = 2.8, R$_{wp}$ = 8.1, R$_{exp}$ = 4.8). Measured data are black crosses and the calculated pattern is the red solid line. The difference between the measured and calculated patterns is shown as the green line and vertical bars are the reflection markers for TiN$_x$ (top) and Ti$_2$AlN (bottom). Note that the unindexed weak reflection at ~28° arises from λ/2 contamination in the incident beam, the regions with contaminations are excluded in (c) and (d) during refinement.

Activation Energy

The relative phase abundances during the isothermal decompositions of Ti$_4$AlN$_3$ at 1400, 1450, 1500, and 1550 °C as a function of time were fitted with a 4th order polynomial function [$y(t)=At^4+Bt^3+Ct^2+Dt+E$] (See Fig. 6). We note that this function is a phenomenological model only. Using the derivative of the fitted functions allows the reaction rate functions to be determined, in the form of the 3rd order polynomial function [$y'(t)=4At^3+3Bt^2+2Ct+D$]. As the reaction rate changes with time, the mean value of the reaction rate function was derived and used as a representative rate constant for the temperature. The mean value of the reaction rate can be obtained using:

$$Mean = \frac{1}{b-a}\int_a^b y'(t)\cdot dt = \frac{1}{b-a}\left[y(t)\right]_a^b = \frac{1}{b-a}\left[y(b)-y(a)\right] \qquad (1a)$$

$$Mean = \frac{1}{b-a}\left[\left(Ab^4+Bb^3+Cb^2+Db\right)-\left(Aa^4+Ba^3+Ca^2+Da\right)\right] \qquad (1b)$$

When a = 0, the equation becomes:

$$Mean = \frac{1}{b}\left(Ab^4+Bb^3+Cb^2+Db\right) \qquad (1c)$$

where A, B, C, and D are the coefficients of the polynomial function, and b is the total time for the isothermal hold.

Table 1 lists the coefficients (A, B, C, and D) and the average reaction rate constant determined for Ti$_4$AlN$_3$ at the various temperatures. The profile fits for the isothermal decomposition at 1400, 1450, 1500, and 1550 °C with the polynomial function are shown in Fig. 6. Applying the same method to Ti$_2$AlN, the change in the relative phase abundances in Ti$_2$AlN at 1450, 1500, and 1600 °C were plotted as functions of time and the fits of the 4th order polynomial function are shown in Fig. 7. The coefficients (A, B, C, and D) and the derived average reaction rate constant for Ti$_2$AlN at various temperatures are summarised in Table 2.

Table 1: Summary of coefficients of polynomial function fitting the isothermal decomposition of Ti$_4$AlN$_3$ and the calculated average rate constant for the corresponding temperature.

Temperature (°C)	A	B	C	D	k	Time (min)
1400	4.02×10^{-08}	-4.72×10^{-06}	1.65×10^{-04}	-1.84×10^{-03}	-2.49×10^{-04}	60
1450	-6.41×10^{-10}	1.85×10^{-07}	-1.47×10^{-05}	-5.11×10^{-05}	-2.57×10^{-04}	150
1500	-1.13×10^{-09}	1.73×10^{-07}	-9.71×10^{-08}	-8.84×10^{-04}	-8.20×10^{-04}	150
1550	8.68×10^{-10}	-5.49×10^{-07}	1.17×10^{-04}	-1.07×10^{-02}	-2.57×10^{-03}	150

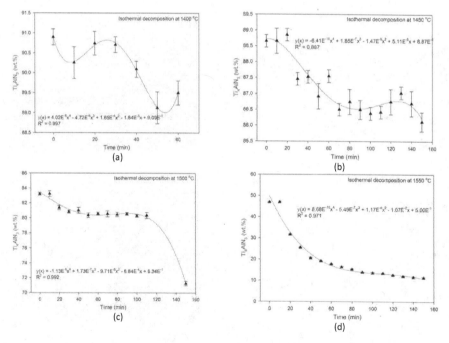

Fig. 6: Phase abundances versus soaking times during the isothermal decomposition of Ti$_4$AlN$_3$ at the temperatures: (a) 1400 °C, (b) 1450 °C, (c) 1500 °C, and (d) 1550 °C.

In order to determine the activation energy for the decomposition of Ti$_4$AlN$_3$ and Ti$_2$AlN, the natural logarithm of k (ln k) was plotted as a function of inverse of temperature (K^{-1}), allowing an activation energy to be derived using the Arrhenius equation (Fig. 8). The negative slope of the fitted lines for vacuum decomposition of Ti$_4$AlN$_3$ and Ti$_2$AlN indicated activation energies of 410.8 ± 50.0 and 573.8 ± 130 kJ/mol, respectively.

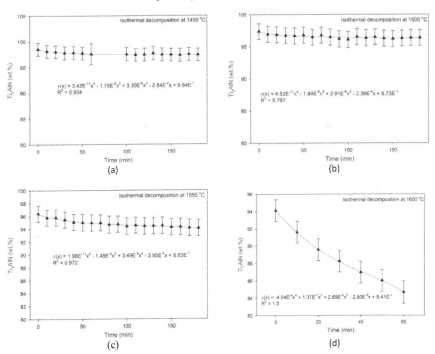

Fig. 7: Phase abundances versus hold times during the isothermal decomposition of Ti$_2$AlN at various temperatures: (a) 1450 °C, (b) 1500 °C, (c) 1550 °C, and (d) 1600 °C [the missing section in Fig.7a is due to a temporary interruption of the neutron beam from the ISIS source.

Table 2: Coefficients of the polynomial function fitted to the data for the isothermal decomposition of Ti$_2$AlN and the calculated average rate constants, at various temperatures.

Temperature (°C)	A	B	C	D	k	Time (min)
1450	3.42×10^{-11}	-1.75×10^{-08}	3.30×10^{-06}	-2.84×10^{-04}	-5.75×10^{-05}	180
1500	4.52×10^{-11}	-1.84×10^{-08}	2.91×10^{-06}	-2.39×10^{-04}	-4.78×10^{-05}	180
1550	1.96×10^{-11}	-1.45×10^{-08}	3.49×10^{-06}	-3.95×10^{-04}	-1.22×10^{-04}	180
1600	-4.04×10^{-09}	1.37×10^{-07}	2.69×10^{-05}	-2.80×10^{-03}	-1.57×10^{-03}	60

The activation energy for the decomposition of Ti$_4$AlN$_3$ is much lower than for Ti$_2$AlN; this may be attributed to its bonding structure. It is proposed that the decomposition of Ti$_4$AlN$_3$ and Ti$_2$AlN occurs by sublimation of Al (mainly) and, to a lesser extent, Ti, from the crystal structure. Although the main bond lengths (e.g. Ti-Al and Al-Al) in Ti$_4$AlN$_3$ and Ti$_2$AlN are similar,[2, 13] Ti$_4$AlN$_3$ has a much larger volume per unit cell, resulting in a less stable crystal structure of

Ti$_4$AlN$_3$ compared to Ti$_2$AlN. This is consistent with the lower energy required to decompose Ti$_4$AlN$_3$ than Ti$_2$AlN. Moreover, as revealed in our previous work for ternary carbides,[14-16] the pore sizes of the surface-decomposed microstructure enable the sublimation of Ti and Al to progress with a minimum resistance and thus increase the rate of decomposition with temperature, and therefore play a critical role in determining the rate constant and the activation energy for the decomposition process. A closer examination of the microstructure of decomposed Ti$_4$AlN$_3$ and Ti$_2$AlN, shown in Fig. 9, is necessary to elucidate the decomposition mechanism. Micro-pores (> 2μm), which can be observed on the TiN$_x$ surface layer of vacuum decomposed Ti$_4$AlN$_3$, are rarely seen in decomposed Ti$_2$AlN. Therefore, the sublimation of Al from Ti$_4$AlN$_3$ is easier than from Ti$_2$AlN, resulting in a faster decomposition rate. [17-18]

Both Ti$_4$AlN$_3$ and Ti$_2$AlN are constructed from a closed-packed Al sheet sandwiched between the twinned Ti$_6$N layers, whereas Ti$_2$AlN is constructed from a single layer and Ti$_4$AlN$_3$ from a triple layer (Fig. 10). The Al-Ti bond lengths are similar in Ti$_4$AlN$_3$ and Ti$_2$AlN, but the distance between the Al and N, between the twinned Ti$_6$N layers, in Ti$_4$AlN$_3$ is much longer than that in Ti$_2$AlN. This structure decreases the stability of the compound.

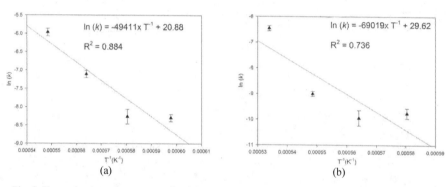

(a)   (b)

Fig. 8: Determination of apparent activation energy for the high-temperature decomposition of (a) Ti$_4$AlN$_3$ and (b) Ti$_2$AlN in vacuum for a range of temperatures between 1400 to 1550 °C.

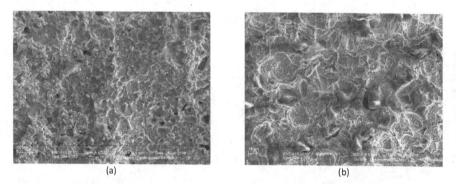

(a)   (b)

Fig. 9: Scanning electron micrographs showing the micro-structural features of (a) Ti$_4$AlN$_3$, and (b) Ti$_2$AlN, decomposed at 1600 °C.

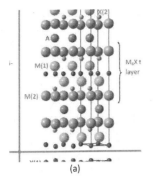 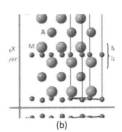

(a) (b)

Fig. 10: Crystal structure of (a) Ti_4AlN_3 and (b) Ti_2AlN.

Compared to the reported activation energies for Ti_3SiC_2 (179.3 kJ/mol), Ti_3AlC_2 (-71.9 kJ/mol), and Ti_2AlC (85.7 kJ/mol),[14-16] the activation energy for the thermal decomposition under vacuum of the ternary nitrides investigated here are much higher. We attribute this to the importance of temperature to the decomposition of Ti_4AlN_3 and Ti_2AlN. For example, at temperatures higher than 1550 °C we measure a relatively larger decomposition rate, incurring a much larger slope of the Arrhenius fit, compared with that observed at lower temperatures. In addition, the microstructure of the TiN_x surface layer formed during the vacuum decomposition is less porous for both Ti_4AlN_3 and Ti_2AlN, compared with Ti_3SiC_2 and Ti_3AlC_2. The calculated activation energies for both Ti_4AlN_3 and Ti_2AlN decomposition are therefore consistent with this theory.

CONCLUSIONS

A new method has been proposed for determining the activation energy of the phase decomposition of Ti_4AlN_3 and Ti_2AlN at temperatures up to 1600 °C. Ti_4AlN_3 was observed to undergo decomposition above 1400 °C, but during the decomposition of Ti_2AlN above 1500 °C we observe the formation of a porous surface layer of TiN_x via the sublimation of high vapour-pressure Al (mainly) and Ti. The kinetics of the isothermal phase decomposition were modelled using a 4th order polynomial function and the reaction rates at these temperatures were determined. The activation energy for the decomposition of Ti_4AlN_3 and Ti_2AlN in vacuum was determined using Arrhenius' equation and found to be 410.8 ± 50.0 and 573.8 ± 130 kJ/mol, respectively.

ACKNOWLEDGEMENTS

This work was funded by the Australian Research Council through a Discovery-Project grant (DP0664586), a Linkage-International grant (LX0774743) and a LIEF grant (LE0882725). The collection of neutron diffraction data was conducted at ISIS (RB1010064) which was provided by the Science and Technology Facilities Council, with financial support from an AMRFP grant. The HRPD (Echidna) data was collected at OPAL with funding from AINSE and the Bragg Institute (P1431).

REFERENCES
[1]M.W. Barsoum, & T. El-Raghy, The MAX Phases: Unique New Carbide and Nitride Materials. *American Scientist,* 89, 334-343 (2001).
[2]C.J. Rawn, M.W. Barsoum, T. El-Raghy, A. Procipio, C.M. Hoffmann & C.R. Hubbard, Structure of Ti_4AlN_3 - A Layered $M_{n+1}AX_n$ Nitride. *Materials Research Bulletin,* 35, 1785-1796 (2000).

[3]R. Pampuch, Advanced HT Ceramic Materials via Solid Combustion. *Journal of the European Ceramic Society*, 19, 2395-2404 (1999).

[4]Z.M. Sun, Progress in Research and Development on the MAX Phases -a Family of Layered Metallic Ceramics. *International Materials Reviews*. In press.

[5]Z.J. Lin, M.J. Zhuo, M.S. Li, J.Y. Wang & Y.C. Zhou, Synthesis and Microstructure of Layered-ternary Ti$_2$AlN Ceramic. *Scripta Materialia*, 56, 1115-1118 (2007).

[6]P.O.A. Persson, S. Kodambaka, I. Petrov & L. Hultman, Epitaxial Ti$_2$AlN(0001) Thin Film Deposition by Dual-Target Reactive Magnetron Sputtering. *Acta Materialia*, 55, 4401-4407 (2007).

[7]M. Yan, Y.L. Chen, B.C. Mei & J.Q. Zhu, Synthesis of High-purity Ti$_2$AlN Ceramic by Hot Pressing. *Transactions of Nonferrous Metals Society of China*, 18, 82-85 (2008).

[8]M. Yan, B. Mei, J. Zhu, C. Tian & P. Wang, Synthesis of High-purity Bulk Ti$_2$AlN by Spark Plasma Sintering (SPS). *Ceramics International*, 34, 1439-1442 (2008).

[9]J.H. Sharp, G.W. Brindley & B.N. Narahari Achar, *Journal of the American Ceramic Society*, 49, 379-382 (1966).

[10]J.D. Hancock & J.H. Sharp, *Journal of the American Ceramic Society*. 55, 74-77 (1972).

[11]A.K. Galwey & M.E. Brown, Application of the Arrhenius Equation to Solid State Kinetics: Can this be Justified? *Thermochimica Acta*, 386, 91-98 (2002).

[12]J. Rodriguez-Carvajal & T. Roisnel, FullProf 98 and WinPLOTR New Windows 95/NT Applications for Diffraction. *International Union for Crystallography, Newsletter No. 20 (May-August) Summer* (1998)

[13]W. Jeitschko, H. Nowotny & F. Benesovsky, Ti$_2$AlN, eine stickstoffhaltige H-Phase. *Monatshefte fuer Chemie*, 94, 1198-1200 (1963).

[14]W.K. Pang, I.M. Low, B.H. O'Connor, A.J. Studer, V.K. Peterson, Z.M. Sun & J.P. Palmquist, Comparison of Thermal Stability in *MAX* 211 and 312 Phases. *Journal of Physics: Conference Series*, 251, 012025 (2010).

[15]W.K. Pang, I.M. Low & Z.M. Sun, *In-Situ* High-temperature Diffraction Study of Thermal Dissociation of Ti$_3$AlC$_2$ in Vacuum. *Journal of the American Ceramic Society*, 93, 2871-2876 (2009).

[16]W.K. Pang, I.M. Low, B. H. O'Connor, A.J. Studer, V.K. Peterson & J.P. Palmquist, In-Situ Diffraction Study of Thermal Decomposition in Maxthal Ti$_2$AlC. *Journal of Alloys and Compounds*, 509, 172-176 (2011).

[17]W.K. Pang, I.M. Low, S.J. Kennedy & R.I. Smith, *In-Situ* Diffraction Study on Decomposition of Ti$_2$AlN at 1500-1800 °C in Vacuum. *Materials Science & Engineering A*, 528, 137-142 (2010).

[18]I.M. Low, W.K. Pang, S.J. Kennedy & R.I. Smith, Study of High-temperature Thermal Stability of *MAX* Phases in Vacuum. *Journal of the European Ceramic Society*, 31, 159-166 (2011).

Author Index

Author Index